The Secret Physics of Coincidence

Rolf Froböse studied chemistry. Following his studies he worked as a research assistant at the Max Planck Society, was division manager at the technology magazine *highTech* and chief editor of the journals *Chemische Industrie* and *Europa Chemie*. Since 1995 he has worked as a freelance science and economic journalist, reporting on research and development issues. He has also penned numerous popular non-fiction books, including *Lust und Liebe – alles nur Chemie? (Lust and Love – is it more than Chemistry?)*, which he authored in collaboration with his wife Gabriele. The latter has so far been translated into English, Spanish, Danish and Korean.

Rolf Froböse

The Secret Physics of Coincidence

Quantum phenomena and fate –
Can quantum physics explain
paranormal phenomena?

Bibliographic information at the German National Library
This publication has been recorded by the German National Library in the German National Bibliography; detailed bibliographic data can be found on the Internet at http://dnb.d-nb.de.

© 2012 Rolf Froböse
Printing, Production and Layout:
BoD – Books on Demand
ISBN 978-3-8482-3445-5

Table of Contents

Preface and Introduction	11
What a Coincidence!	14
The Lost License Plate	14
People Think – And God Steers	16
Twins Meet after 14 Years…	17
…And In the Marital Bed	18
Three Siblings with the Same Date of Birth	18
From Spies and the Titanic	19
Can that Be a Coincidence? What Secretly Binds John F. Kennedy and Abraham Lincoln Together	20
The Magical Number 273	24
September 11, 2001 and the Apparent Magic of the Number Eleven	25
The Dialog, Part 1: From Number Myths, First-Order	27
The Major Who Wiped Away a Tear	31
Coincidences to Manipulations	27
Coincidence and the "Gretchen Question"	32
The Cousin	32
The Appearance of a Deceased Wife in a Puzzle and Other Curious Occurrences	34
Coincidence or Industrial Espionage?	36
Of Lost Wedding Bands, Dentures and Steinway Grands	36
Coincidence or Telepathy?	39
Mr. Froböse, You Are under Arrest!	40
"Inspector Chance" in Science	42
A Strange Dream Leads to a Fundamental Discovery	44

How a Chemists' Forgetfulness Helped Create a Super Substance   47

Can that Still Be Considered Coincidence or Is There a Hidden Higher Order Behind It?   49
- The Family Crest   49
- Zuckmayer's Curious Wallpaper   50
- The Mysterious Double Exposure   51
- It Doesn't Always Have to Be Plum Pudding   51

The Dialog, Part 2: Some Doubt Arises   52
- Why Psychologist and Psychiatrist C. G. Jung Was So Fascinated by Coincidence   53
- The Discovery of the Periodic System of Elements – A Case of Synchronicity?   54
- A Dream Becomes Reality   56
- Beatles Forever   57
- Death of a Newscaster   58
- The Peculiar Affairs Had by Ann   58
- The World Is Quite Literally Small   59
- A Grandfather's Mysterious Inspiration   60
- What Gave an IT Expert Goosebumps   60
- "Preliminary Round" of a Class Reunion   61
- Greetings from Venus   62
- An Odd Coincidence on Christ the King Sunday   64
- Pseudonyms and Reality   66
- A Church Choir's Guardian Angel   67

First Physical Interpretations of the Phenomenon   68
- How the Soul Became Taboo No Longer   69

The Dialog, Part 3: Coincidence Does Not Always Equal Coincidence   71

Spooky Physics: A Short Excursion into Quantum Mechanics   73
- Almost Esoteric: Strange Action at a Distance   74

Will We Someday Be Able to Travel by Teleportation?	76
From Schrödinger's Cat to the Many Worlds Theory and Quantum Immortality	78
Strange Encounters in Student Housing	81
Tragic Death of a Classmate	82
The Grandma from the Afterworld	82
The Dialog, Part 4: Our Dual Ego	84
Father Predicts His Death	85
The Scientist's Secret Fear of the Unknown	87
Is the Other Side the Great Internet of Reality?	90
The Dialog, Part 5: Where Is God?	93
The Soldier who Said Goodbye	94
"Willy – Where on Earth Did You Come From?"	95
The Dialog, Part 6: Why There Can Be No Stable Channels of Communication on the Other Side	96
"Seeing" a Major Fire from 450 Kilometers Away	98
From the Primordial Soup to DNA	99
The Dialog, Part 7: A Series of Conclusions	103
Further Reading	107

Dedication

This book is dedicated to my mother-in-law, Waltraut Heubach, who died peacefully in her bed during the night from 14th to 15th July 2012 in Seesen/Harz, in Northern Germany. It was just the way she had always wished for.

Dr. Rolf Froböse
Wasserburg/Inn, December 2012

Preface and Introduction

Let's begin with a question for the reader: you have surely encountered the infamous "Inspector Chance" at one time or other in your life, have you not? And when something unexpected suddenly interrupted our supposedly "perfect little world" that so depends on reliability, order and predictability, you probably yelled or at least thought to yourself, "That's not possible!"

That kind of reaction is completely normal, as coincidence generally does show up unexpectedly. While it is certainly welcome if we are winning the lottery, we tend to argue with fate when a dramatic accident occurs, and in the field of science we are fairly certain that there can be no real coincidence. Is coincidence then "really" coincidence or simply outside the field of our subjective vision? Does the unexpected only seem to be coincidental because we fail to recognize the structure behind it? And if there is a structure, is our reaction to what has happened to us not also pre-defined? Is everything thus our inescapable fate?

"Our brain is programmed to identify connections and causes." This thesis, proposed by more than a few psychologists, requires no further explanation. It is the only way we learn to understand the world from early childhood on. This property explains why we tend to attribute to certain events a deeper meaning, even though they are simply and poignantly "coincidental".

A typical example is a call from a friend we were just thinking about and from whom we had not heard in a long time. Many people assume there must be a deeper connection. Yet, they forget that we often think of someone and are not called by them. Another example is an innocent person who ends up destitute and who is so desperate they no longer know up from down. Yet, he wins the lottery over the weekend and his entire life changes. "That can't be coincidence," he would say, because he affiliates the event with his own situation and attributes a higher meaning to it.

"The chance of getting six right in the lottery is around 1 in 14 million," a mathematician would predict. However, our winner could never be satisfied with a mere statistic. "But why me of all people? There must be more to it!" This kind of reaction is completely understandable; however, even when the chance is 1 in 14 million, the chance of at least one big win increases when millions of people are playing. Is the unpredictable really merely coincidence and a question of mathematics as some claim?

Psychologist and psychiatrist Carl Gustav Jung, a student of Sigmund Freud, did not believe in coincidence. Instead, he assumed there was a higher order behind the phenomenon that steered our lives. He referred to this order as synchronicity. In other words, an observer can still see a series of sequential events that closely follow one another, but are not connected by a causal relationship, as making sense and thus logical. According to Jung, synchronicity is a precursory internal event – a dream, an idea, a vision or even an emotion – that precedes an external event. The external event, in return, acts as a reflection of or an answer to the internal event.

This interpretation was well received by many people, whether John Doe, academics or scientists. "Chance is perhaps the pseudonym of God when he does not wish to sign his work" is how French author and Nobel Prize winner in literature, Anatole France, once formulated it.

Back to our key question: Does coincidence exist or not?

The question cannot be answered with a simple "yes" or "no". The author, however, considers it naïve to write off all events – as unlikely as they may be – as mere coincidence. Conversely, it is just as absurd to search for a higher meaning in absolutely everything – on vacation, on the beach in Rimini, when you happen to run into your neighbor, etc..

Instead, he hopes to show that coincidence has two very different faces. The first face, which the author refers to as first-order coincidence, shows us the trivial side: a reality that can be explained by mathematical rules, but which we allow to fool us time and again, causing us to search for internal cohesion. The second face is the "coincidence of a higher order", which is not actually coincidence any longer as it is based on connections and which scientists have only just begun to identify and to research. The cornerstone of this new branch of science was placed long ago by C. G. Jung when, in collaboration with physicist and Nobel Prize winner Wolfgang Pauli, he searched for physical signs of synchronicity.

The author would like you to join him on a phenomenal journey of discovery through the world of coincidence. Let yourself be surprised at how coincidence has presented itself to him, his friends, relatives and acquaintances in completely different forms. See for yourself that there is no real "Supernature", rather a form of nature that simply has not been discovered yet and what revolutionary ideas modern scientists are already expressing about the phenomena of "coincidence of a higher order". "It is fully plausible that, for example, a picture falling from the wall and the simultaneous death of a loved one are the result of the same physical effect, one that can be derived from the entanglement principle of quantum mechanics," is his courageous summary.

To better understand the sheer unbelievable, the author has set aside two "travel companions" for the reader: Al and Zach, two friends who are physicists and literally as different as A and Z. While Al vehemently defends the "classic" side of physics, Zach belongs to the avant-gardists in his field who approach the phenomena described above with no taboos, and attempt to reconcile it with the discoveries of modern quantum physics.

Let yourself be carried away – not only while reading these incredible stories of coincidence, but by the exciting discussion as well, which is free of prejudice and does not attempt to place mysticism on a scientific platform.

Let the journey begin!

What a Coincidence!

"Coincidence is the only legitimate master of the universe."
(Napoleon I. Bonaparte)

Coincidences accompany us at every step: at work, during our free time, on vacation. Looking at it from that point of view, there is no situation – however far off the beaten path it may be – that can protect us from the unexpected. I truly became aware of that fact as I was writing off a few odd coincidences I had come across during my life. My theory was strengthened by all of the strange stories other people had told me. A few cases, which I will share with you here, I found in media reports or other sources available to the public. But let's get to the point:

The Lost License Plate

The first example describes how a lost object made its way back to its rightful owner. What is astonishing in this case, is that there was an incredible connection between the finder and the owner. The author himself is the focal point of this story.

It started when I was out walking our white German shepherd dog "Aragorn" on October 14, 2006. We were walking through the meadow near the lake in Reitmehring when, to my great frustration, I discovered that the front license plate on our car was missing. I consulted my wife and she, like me, assumed we were dealing with a theft. So I reported it to the police. The day was a Saturday.

On Monday, October 16, the telephone rang at 7.20 a.m., quite an unusual time for us. A caller wanted to know if there was anything missing from my car.

"Don't tell me you found a license plate that comes to the name of RO-GF…?"

"Yes, that's what I have right here. I found it on Saturday in front of the cemetery in Edling."

"Yes, yes, my wife went to a funeral on Friday," I answered.

"Well," answered the caller, "then there are three odd coincidences."

"And those would be?"

"First, I am the head of the vehicle registration office in Rosenheim. Second, you don't have to drive all the way out to Rosenheim because I live in the Reitmehring neighborhood of Wasserburg, just like you. If you'd like to stop this evening after 7.00 p.m., you can pick up your license plate." He gave me his address and I was glad I basically only had to go "right around the corner" and not drive 20 miles to Rosenheim.

"And now for coincidence number three," he continued and let it hang in the air for a minute to torture me. He then asked, "Mr. Froböse, where were you born and where did you grow up?" Of course, he already knew the answer. As the head of the vehicle registration office he had looked up my information on the computer over the weekend.

"In Seesen at the Harz mountains," came my proper reply. "And that is where I am from as well!" - I almost fell off my chair and began to look forward to our meeting that evening.

Well – the person who opened the door was a man whose face seemed very familiar. We exchanged memories over a beer and realized that we had both lived in Garden Street until 1956, not even 30 meters apart. Our houses had been kitty corner to one another and we had even gone to the same school. Neither one of us could believe that a license plate lost in a town 400 miles from Seesen at the Harz mountains with only 1,000 inhabitants in the entire town of Reitmehring had brought us together. Not to mention the fact that the meadow near the lake in Reitmehring, where I tend to park my car when I take the dog out, was only 30 meters from the finder's house.

Another coincidental event occurred that was along the same lines. It is a story that a Web designer named Christian Gruber living in the upper Bavarian town of Geretsried told me not all that long ago: "After not seeing a childhood friend from the neighborhood I grew up in for more than 20 years, as we both had changed locations and our professional paths had taken us in different directions, I ran into his sister and her husband and children in Bulgaria this summer. We were even staying at the same hotel, on the same floor, just three rooms apart! I had often asked myself what my old friend was up to nowadays. So his sister and I exchanged addresses at the hotel and now I am back in contact with my old friend. After having both moved repeatedly during that time, he only lives about 150 meters as the crow flies from me with his own family."

Another example comes from engineer Jochen A. Wanderer from the town of Heere near Salzgitter, who also ran into an old acquaintance fully unexpectedly.

People Think – And God Steers

"I have been dealing with the phenomenon of coincidence for about 30 years. It was the spring of 1977 and the trade fair in Hanover was about to begin as it does every year. At the time I was working in the Documentation & Editing Department at GDMB (Association of German Steel Workers and Miners) in Clausthal-Zellerfeld as a student worker. I was studying metallurgy and had been with the company for two users. My boss, the assistant to the mining engineer Herby Aly, and I made the drive from Clausthal to Hanover. An avalanche of traffic from all of the highways accumulated in front of the trade fair parking lot with traffic from the A2 that runs between Berlin and Dortmund, the A7 that runs between Hamburg and Kassel, as well as all of the local and regional traffic from the B6 highway. The shortcut to the trade fair and all of the entrances to the parking lots were – as they have been every year – completely backed up."

Aly laughed when I commented on how even the "mega managers" in their luxury cars were at their wits end and had delegated the management to the parking lot attendants. There were two lanes of traffic coming together from different directions, one with traffic from the A2 coming out of the Ruhr area and the other, like us, from the A7 coming from the Harz, and cars were let in alternately. As I was getting out of the car, I was very surprised to see an old friend from college, who by now had his PhD, pull in from the other lane and park right next to us. Sitting there in his fat old Mercedes-Benz was Dr. Thomas Peters. Everyone was astonished at the coincidence – we laughed and crossed over the bridge to the trade fair together – but we have never seen each other since…was it really just a coincidence?"

Yet another story, which is even more incredible, was appeared on "stern.de" on July 13, 2007:

Twins Meet after 14 Years…

Just imagine you walk into a restaurant and see someone who looks almost exactly like you. That is precisely what happened to two 14-year-old girls in July 2007 in Ecuador. For 14 years neither knew of the existence of her twin. Now they were both overjoyed.

According to medical reports, an Ecuadorian doctor concealed the fact that his patient was pregnant with twins and kept one of the girls for himself when they were born. The case was not exposed until 14 years later when friends of one of the girls, who lived in the same area near the city of Guayaquil, told her about a look-alike and arranged a meeting. "We hugged each other and were moved to tears," the newspaper *El Comercio* quoted Andrea, the twin who had grown up with her biological parents.

The biological parents have now sued for custody. "They never told us that we actually had twins," said the father Augusto Freire. Although their "new" daughter, Marielis, was thrilled about her sister, she wished to remain with the doctor and his wife, the newspaper continued. The doctor accused the biological parents of having left the second child at the hospital back in 1992. The baby had to be treated at a different clinic for a few days after she was born due to health concerns and by the time she was returned to the clinic in which she was born, the biological parents had disappeared. The Public Attorney's Office has begun investigating the matter.

...And In the Marital Bed

Coincidence played a nasty joke on a young couple in Bulgaria. The Berliner Kurier newspaper reported the following unbelievable story on August 22, 1997. "To their horror, a pair of newly weds in the Bulgarian city of Plovdiv found out they were twins. As a result, they submitted divorce papers only a week after they were married. Marietta and Wenelin Wassilewi grew up in the same home, but were adopted by different couples. Mariette and Wenelin met in 1995 in the Black Sea city of Varna. Shortly after their marriage, the husband wished to introduce his bride to his birth mother. Upon realizing the newly weds were her twins, she fainted."

Three Siblings with the Same Date of Birth

The Austrian daily newspaper "Die Krone" published a recent example:
"The 2nd of October 2007 became a fateful day for the Cotton family of Marysville in the US State of Ohio: Jenna Cotton gave birth to a child on this date three times in a row. The birthday parties for four-year-old Ayden and his little brother Logan had just ended last Tuesday when their extremely pregnant mother gave birth to her third child that same evening, reported US magazine *Columbus Dispatch* on Friday. According to calculations performed by scientists at the University of Ohio, the statistical probability of a woman giving birth to three children on the very same day in three different years is around 7.5 in a million claimed the report. A C-section was not necessary for any of the births."

The author had to smile at one section of the report. Was it really necessary for scientists to calculate the statistical probability? After all, the likelihood of a second child being born to a family on the same day as its first child is quite simply 1 in 365. The probability of three siblings being born on the same day is thus reduced to 1 in 365 x 365. According to Adam Riese that produces a total probability of 1 in 133,225 or 7.5 in a million. Jane Doe from the supermarket could have told you that!

I have provided a small collection of stories of coincidence that I collected from a variety of sources to conclude this first chapter:

From Spies and the Titanic

Two vehicles collided in Worcester, England in October 1974. Neither driver was injured. After exiting their vehicles, they exchanged business cards. They realized they both shared the same first name of Frederic and the same last name of Chance.

When American author Hart Crane composed a poem about the Brooklyn Bridge, he was unknowingly living in the former apartment of Colonel Washington Roebling. Roebling was the Chief Engineer who oversaw the construction of the Brooklyn Bridge.

While author Normal Mailer was working on his book *Barbary Shore*, he was suddenly overcome by the feeling that he needed another Russian spy. He first introduced the character into his book in a supporting role. Yet, during his writing, the invented spy continuously "forced" himself to the forefront. In the end, the entire story was about the spy. After the book was published, the FBI stormed Norman's house and arrested a man in the flat below who turned out to be Colonel Abel of the Soviet KGB – a top spy.

A novel by Morgan Robertson entitled *Futility* was published in 1898. The book described how a ship crashed against an iceberg in the North Atlantic. The author named this fictional ship the "Titan". A real ship crashed into an iceberg at that exact location in the North Atlantic in 1912, the famous "Titanic". The same thing happened to another ship in 1939, a ship called "Titanian".

In *The Narrative of Arthur Gordon Pym*, author Edgar Allen Poe described how, in their desperation, three shipwrecked sailors killed and ate a cabin boy named Richard Parker. 50 years later, three shipwrecked sailors were actually forced in front of a court of law under charges of cannibalism, having devoured a cabin boy named Richard Parker.

Two Taiwanese, Lee Hsi-lung and Lai Yun-kun, were both born on June 25, 1973 in the city of Taitung. Neither had ever met before attending a party hosted by a mutual friend. Unfortunately, there was not much time for them to become acquainted as the car in which they were sitting was hit by a truck. Both died just a few meters from the place where they had been born 19 years before.

During World War II, a renowned American photographer took a picture of a field in which lay a dead GI named Tannenbaum, meaning Christmas tree. When he returned to that spot years later, he realized that a Christmas tree farm had been planted in this field.

Can that Be a Coincidence? What Secretly Binds John F. Kennedy and Abraham Lincoln Together

In April 2001, the author and his family spent some time in Texas. The first stop on our trip was Dallas. From our room on one of the upper floors of the Hyatt Hotel, we enjoyed a view of the city's skyline. As we found out later, we were also able to see the J. F. Kennedy Museum about 1.5 miles away. The museum is located in the warehouse from which Lee Harvey Oswald fired the deadly shot at the president in 1961.

The visit to the museum was naturally a "must" for us. In addition to numerous details of the circumstances surrounding the assassination, including countless pictures and newspaper reports, there was also no shortage of conspiracy theories. Yet, what also interested us was a very odd, almost uncanny aspect.

The aspect in question was a truly stupefying parallel that connected the fates of John F. Kennedy and Abraham Lincoln. "This duplicity of events is a clear sign that coincidences are not always random, but can probably befall people as part of their fate," states one popular theory. Or is there something "fishy" about the story itself? Before I address this question, I would like to present a series of strange parallels:

- Both President Lincoln and President Kennedy addressed the issue of civil rights and African Americans and were thus shunned by the South.
- Lincoln was elected to congress in 1848 and Kennedy was elected to congress in 1948.
- Lincoln was elected President in 1860 and Kennedy was elected President in 1960.
- Both knew a certain Dr. Charles Taft. Lincoln was treated by Dr. Charles Sabin Taft, Kennedy knew Dr. Charles Phelps Taft, the son of President William Howard Taft.
- Both had a friend named Graham. Lincoln's friend's name was William Mentor Graham, a teacher from New Salem (Illinois) and Kennedy's friend's name was Billy Graham.
- Both had a friend named Adlai E. Stevenson, who was a democrat from Illinois. Lincoln's friend became vice president to Grover Cleveland and Kennedy's friend was a two-time presidential nominee for the Democratic Party.
- Lincoln had two sons named Robert and Edward. Kennedy had two brothers named Robert and Edward.
- Both were tall and athletic, loved rocking chairs, had good senses of humor, quoted Shakespeare and the Bible and suffered from eye muscle paralysis, which was distracting at times. Both were boat captains.

- Neither was afraid of death, which is why they categorically refused body guards. Yet both of them spoke time and again of how easy it would be to murder a president.
- In 1861 a police inspector named John Kennedy thwarted an attempt on President Lincoln's life.
- The successors of both presidents were named Johnson, were Democrats from the South and were members of the Senate. Both were tall, had two daughters and were the only presidents to suffer from urinary stones. Neither trusted their predecessor's cabinet. Andrew Johnson was born in 1808, Lyndon B. Johnson in 1908.
- Lincoln's alleged murderer, John Wilkes Booth, was born in either 1838 or 1839 (depending on the source). Kennedy's alleged assassin, Lee Harvey Oswald, was born in 1939.
- Both assassins were murdered before their trials could begin.
- Investigations into whether their murders were a conspiracy were performed for both presidents. In both cases, investigations were resumed after many years without being able to explain with certainty who all was involved.
- Both presidents were killed on a Friday as they sat next to their wives.
- Both presidents were accompanied by another couple and although both men were injured during the attack, but did not die immediately, the women were not hurt.
- Both were shot in the head from behind.
- Lincoln was killed in Box 7 of the Ford Theater. Kennedy was killed in a Ford Lincoln, which was the 7th vehicle in a convoy.
- Both were autopsied by the military, interred in mahogany coffins and were placed at rest in the same catafalque.
- Both were named after their grandfathers and were the second born.
- Both had a sister die prior to their presidential election.
- Both married dark-haired, 24-year-old women, who had been engaged before.

- Both women spoke French fluently and belonged to the prominent social class.
- Both women had the White House renovated after many years of neglect.
- Both had four children, two of whom died before they were ten years old.
- Both lost a child while they lived in the White House.
- Both the Lincoln and Kennedy children rode ponies around the White House grounds.
- Lincoln's secretary's first name was John; Kennedy's secretary's name was Lincoln.
- One of Lincoln's employees named Kennedy advised the president against going to the theater. One of Kennedy's employees named Evelyn Norton Lincoln advised the president against the trip to Dallas.
- John Wilkes Booth shot Lincoln in a theater and fled to a warehouse.
- Lee Harvey Oswald shot Kennedy from a warehouse and fled to a movie theater.
- Lincoln stopped at a small town in Maryland named Monroe a week before his murder. Kennedy, in turn, had an affair with Marilyn Monroe.
- Following the death of their fathers, Robert T. Lincoln and John F. Kennedy Jr. moved to a new home with his mother and brother / sister, the address of which was 3014 N Street NW, Georgetown and 3017 N Street NW, Georgetown, respectively.

My relative Jochen A. Wanderer from Heere brought my attention to another oddity, namely:

The Magical Number 273

Which complex meaning can we find in nature when we consider the following list of data related to the number sequence 273?

- A pregnancy lasts 9 months or 273 days.
- It takes 27.32 days for the moon to complete a full rotation.
- The reciprocal value of a leap year is $1 / 366 = 0.002732$ days
- The absolute zero temperature on the Celsius scale is - 273.2 °C.
- 4 divided by ϖ is 1.2732.
- The expansion/reduction of gas is $1 / 273.2$ per °C
- A circle with the same circumference as a square with an edge length of 2 units of length has a radius of 1.2732.
- It takes the sun an average of 27.3 days to rotate around its own axis.
- The acceleration of the sun is 273 m / s².
- The acceleration of the moon along its path around the Earth is 0.273 cm / s².

J. Wanderer commented: "My professor at university, Prof. Ulrich Kuxmann, once hosted a seminar on metallurgy that was attended by six hand-picked students. What he said was that you cannot find the right answer if you are not asking the right question. Is there really no answer to the phenomena of coincidence?"

Another oddity - with various versions are "floating" around the internet - pertains to the events of September 11, 2001. I do not wish to keep this story from the reader, but will only address it in detail in the subsequent chapter "Dialog".

September 11, 2001 and the Apparent Magic of the Number Eleven

- The word Afghanistan has 11 letters.
- The name of the terrorist who threatened to destroy the Twin Towers back in 1993, Ramsin Yuseb, consists of 11 letters.
- The name of the American President George W. Bush also has 11 letters.
- This might all just be a strange coincidence, but it gets even better: New York was the 11th state to join the USA.
- The first airplane that flew into the Towers was Flight Number 11. The flight had 92 passengers. $9 + 2 = 11$.
- The airplane used in Flight Number 77, which also flew into the Twin Towers had 65 passengers. $6 + 5 = 11$.
- This tragedy took place on September 11. Or, as it is referred to today, 9/11. - $9 + 1 + 1 = 11$.
- The date corresponds to the emergency telephone number in America: 911. $9 + 1 + 1 = 11$.
- The total number of all victims who died as a result of the hijacked airplanes was 254. And yet again, $2 + 5 + 4 = 11$.
- September 11 is the 254th day of the year. Again, $2 + 5 + 4 = 11$.
- The bombing in Madrid took place on 3/11/2004. The checksum of that date is $3 + 1 + 1 + 2 + 4 = 11$.
- The tragedy in Madrid took place exactly 911 days after the attack on the World Trade Center. Again 911, again 9 / 11 and again $9 + 1 + 1 = 11$.

And here is where things get really strange:

- Open a Word document and perform the following steps:
- In capital letters type Q33 NY (that is the number of the flight that first flew into the Twin Towers).
- Highlight Q33 NY. Change the font size to 48 and change the font itself to Wingdings.

Go ahead and try it for yourself, but I will do it here for you so that you, dear reader, do not have to put your book down to turn on your computer:

Q33 NY

Changing the font to Windings produces the following result:

I have to admit: I felt a little sick the first time I performed this conversion. I have a more relaxed view of it all today. You will find out why and much more in the following chapter.

The Dialog, Part 1: From Number Myths, First-Order Coincidences to Manipulations

Zach didn't even have to check his watch when the doorbell rang. His friend Al was always exactly on time, down to the minute, and had never understood the much-cited academic "quarter hour" during his studies at the Physics Institute at the University of Göttingen.

"I'm pleased to see you, Al."

"The pleasure is all mine, you have promised me quite an exciting evening. I read the materials you sent me with great interest."

"Yes – the phenomena of coincidence is a subject that is very close to my heart. I would love to hear your opinion on it."

Al raises his eyebrows: "Ha! I knew it! And if I know you at all, I bet you probably not only have a few facts, but a theory prepared as well that you want me to rack my brain over. But don't worry, I actually have enough time today to refute everything."

Zach had to laugh: "That's what I like to hear, you are still the same old Al."

He asks his guests to take a seat and fills their glasses with wine. "What sort of wine is that?"

"A wine from the Lake Constance region of Baden. A dear relative that I ran into quite by accident gave it to me."

The men toast their glasses. "Okay, let me have it," Al encourages his host.

"Let's begin with the last example. What do you think about it?" Zach questioningly raises his eyebrows.

"You mean the matter of September 11, 2001 and the magical number 11?"

"That's exactly what I'm referring to!"

"Well, I find this eleven comparison to be totally far-fetched. And I've never heard of this strange Wingding font."

"That doesn't surprise me. What program are you using on your computer?"

"Windows 2000, why?"

"And which font?"

"Always Times New Roman, I don't care about any of the others."

"Well, if you ever bothered clicking the font, you would find dozens more, including Wingdings."

"What am I supposed to do with a font I can't read?"

Zach can't suppress his grin: "For example, you could write your girlfriend an encoded letter that she wouldn't be able to open until she had converted it. Wingdings is a symbolic font that has around 200 characters."

"And if you enter the flight number using that symbolism you get a very uncanny series of characters with an airplane, symbols representing the twin towers, a skull and crossbones, and the Star of David."

"Exactly."

"That is a bit strange. But that has to be a coincidence – or don't you think so?"

"In this case it actually wasn't a coincidence, but a very sick joke. E-mails were sent out after September 11, 2001 in which the number Q33NY, the alleged flight number of the airplane that stormed into the World Trade Center, was supposed to be formatted using the Wingdings font. However, the alleged flight registration number Q33NY doesn't even exit. The two airplanes that crashed into the World Trade Center were American Airlines Flight 11 and United Airlines Flight 175. And neither of those numbers looks suspicious at all in Wingdings."

"Then I am very relieved."

"I simply wanted to use this crass example to show you that we in our search for coincidences and phenomena will run into a few tricksters who use falsified messages to draw attention to themselves."

Al picks up the materials he received from Zach before. "What about the other examples?"

"Let's stick with the numbers for a minute. Although the story with the 273 is serious, I think the combination is simply a coincidence."

"And what about all of the examples?"

"Those are examples of first-order coincidences that I chose intentionally."

"First-order coincidences?" Al looks at his friend questioningly.

"Yes. There are actually coincidences of a higher order, but we will get to those later in the evening. Yet, I will promise you one thing – it gets more and more exciting."

Al seems satisfied. "Okay. And I was beginning to think you thought there was a mysterious connection between Mr. Gruber and Mr. Wanderer. Similar things have happened to me as well. So I would have vetoed that right away."

Zach takes another sip of wine and savors it for a moment. "I would even consider the case of the license plate to be a first-order coincidence, even if, from a pure mathematical point of view, we can assume that the probability is very low."

"And the thing with Lincoln and Kennedy?"

"The Lincoln-Kennedy mystery is referred to as a series of correlations in the two US presidents' biographies that appeared not long after Kennedy's murder in 1963 and has made the rounds through the media time and again ever since - and on the Internet in the meantime as well. Some of the alleged coincidences have, however, been fudged a little. Wikipedia has a detailed collection on the topic."

"Still, it is an impressive list. Even as a skeptic I have to admit that."

"The phenomena that if you take a large enough group of people, you will be able to find two people with a few characteristics in common such as date of birth is referred to in mathematics as the 'date of birth problem'. Add to that the fact that the number of properties that come into question for that kind of list is very high. It includes the day of the week on which the murder took place, the successor's year of birth, the last two numbers of the year they were first elected to Congress, the number of letters in their secretaries' names, etc. So you could ignore all of the non-matching properties and still come up with a considerable list."

"Another manipulation in other words?"

"Yes, however, not as brazen as the one surrounding September 11th. In reaction to the media's play on the "Lincoln-Kennedy mystery", a newspaper called the *Skeptical Inquirer* hosted a Spooky Presidential Coincidences Contest in 1992. Readers were asked to create their own list of parallels between other pairs of presidents. One of the two winners of the contest found an astonishing list of sixteen coincidences between Kennedy and the former Mexican president Alvaro Obregón. Now there's a sorry effort all across the Internet trying to establish coincidences between Napoleon and Hitler."

"As yet, I don't see any differences in our assessments of the situation." Al leans back, relaxed.

Come, let's take a look at the other coincidences I collected for this evening."

The Major Who Wiped Away a Tear

"The more pains a person takes to plan things, the harder he is hit by coincidence." (Friedrich Dürrenmatt)

Another example of a strange coincidence comes from a German psychologist named Ulrich Borsch, who lives in the Italian province of Umbria. He reported an event that befell him during his military service.

"Years ago when I was assigned to El Paso, Texas for further training, I naturally used the occasion to not only get to know the land better, but the people as well.

I was a common frequenter of the American officers' casino, not only for the cards, but because the food was good and cheap. One afternoon I was unable to find a seat and decided to come earlier the next time. I slowly made my way to the exit and found myself on the street, undecided as to what I should do. I suddenly decided to go back in. There had been no change in the availability of the seating.

I looked around and simply sat down at a table occupied by high-ranked officers. There was only a small space left for me. The soldiers sitting at the table moved a little closer together and there was just enough space. I listened interestedly to their conversation. After a short time an older major addressed me and we ended up in a lively conversation. He wanted to know where I came from, which part of Germany. Here is the conversation as I remember it:

"Where are you from?"
"Northern Germany."
"Oh, what city?"
"A small town south of Hanover."
"Tell me the name of it."
"Seesen at the Harz mountains."

When I said the name the man's eye grew large and he placed his hand on my shoulder.

"Really?"

"Yes, of course. I certainly know where I live."

The discussion at the table had quieted as our conversation grew louder. Everyone was looking at us.

What had happened: It quickly came to light that the major in question had lived in my home town during the last days of the war and had spent another four years there. I was astonished. He knew and was familiar with many things from the city and he named people that I knew as well. It ended up being a long night and every once in a while I saw a tear escape from the man's eyes, which he was always quick to wipe away.

Coincidence? No, I don't think so. What had moved me to go there that afternoon? Why did I turn around? Why did I choose to sit to the major when there had been a little more space at other tables?

Coincidence and the "Gretchen Question"

Author and screenwriter Jurek Becker once told me of another odd event: "There are strange coincidences. For example, I once ended up at a bus stop where a young woman was standing, whom I thought was Gretchen Kosanke, a girl who had been in my class at school and whom I had not seen since. I approached her and was about to talk to her when I noticed from up close that it was not her at all. The bus arrives a few minutes later and who gets off? Gretchen Kosanke."

The Cousin

The following example is another from psychologist Ulrich Borsch.

"The last time I saw my cousin was in 1955 in Seesen. Since he is at least 17 years older than I am and has been living in Switzerland, and off and on in India, we were never in contact. I hardly even remembered him.

One day last year, my brother sent me a letter from Seesen that said our cousin had written him and wished to see a few old family photos. This information was just a side note in an e-mail without more information as to where he was living or a telephone number. I read the lines and saved them in my correspondence folder. A few days later, I spontaneously opened the folder while I was writing a report for work. I quickly entered his name in an international telephone book and received some information. Without delay I reached for the phone and called him. He was very surprised. We quickly developed an intense professional collaboration and a new family connection. Coincidence?

No, I don't think so. I was open to something new, to change. I had been thinking and revising my opinion on certain thoughts and my approach to life for quite a while. Thus, I had become more open. No coincidence at all!"

Uli Borsch had thought about the phenomenon of coincidence as well. In an e-mail he wrote the following to the author: "We sometimes make it a little easy on ourselves. Everything or many things we cannot or do not want to explain, we chalk up as coincidence or even as fate. The latter has more bearing and is thus only mentioned in situations where we question the reach of human action and thought. Take the meaning of the word "to befall", for example. In other words, something falls across our path.

One can examine such things from different points of view. I always try to draw upon the causal connection between the cases for assistance. Cause and effect. Seemingly simple. A certain action triggers a certain reaction or another action. As my considerations and observations do not necessarily involve a direct consequence to an action, they seem to be coincidences as we are no longer aware of the cause or have not registered it as such. I am convinced that there are no coincidences. To me it all seems like a time-delayed, logical flow of action not bound by space in the true sense of the word."

The Appearance of a Deceased Wife in a Puzzle and Other Curious Occurrences

- British widower Stuart Spencer had the surprise of his life back in the beginning of 2001 as he was putting together a puzzle that had been manufactured in Germany. When he completed the puzzle, he saw a picture of his deceased wife who had died a few years ago. It was a picture of her from five years ago on an excursion on a German paddle wheeler. Although it was strange, it was obviously meant to be, Spencer told British news agency PA. His oldest daughter had given him the puzzle for his birthday.
- Two soldiers, who were neither related by blood nor in-laws, were brought to the same military hospital during the First World War. They were both from the German region of Silesia, both were serving voluntary tours in a transport company and both were named Franz Richter.
- Could either of the two above-mentioned soldiers have known that a boat sailing the Atlantic had sunk close to the Welsh coast on December 5, 1664 and that only one of the 81 passengers had survived, a man named Hugh Williams? Had these twins-by-name only heard that another ship sank on that same day, December 5th, 120 years later and that the only survivor was once again a man named Hugh Williams, they may not have thought themselves so particularly special after all.
- It is relatively unlikely that someone would actually get all six numbers right in the lottery. What is even more unlikely is being hit by a meteor. Yet, it happened to a woman from Sylacauga, Alabama. In 1954, she was hit by a 3.9 kilogram meteorite as she was lying on the sofa at home. The stone from the infinite expanses of outer space fell through the roof of her house, bounced off the radio and hit her leg before coming to rest in her living room. She came away with a mere bruise.

- Telephone companies are sometimes better than their reputation. At least that is what Leonardo Diaz would attest to. In June of 2002 he was travelling through the Columbian Andes, far from civilization, when he was surprised by a snow storm. When he went to call for help from his cell phone, he saw that he had used up all his minutes. He spent an entire day freezing, waiting for his imminent death. Suddenly, the telephone rang – it was a woman from the phone company's marketing department who wanted to sell him a new product. Diaz was rescued just a few hours later.
- A man named George D. Bryson checked into a hotel in Louisville, Kentucky. Although he had not booked the room in advance, he found a letter in his room addressed to George D. Bryson, Room 307. However, the letter was not even for him – the intended recipient of the letter, a George D. Bryson from Montreal, who had rented the room prior to his arrival, had just left.
- Once, two single mothers in a clinic in Würzburg gave birth to twins on the very same day. They were placed in the same room. As they began to talk, it turned out that all four children had the same father. It was only through this coincidence that the two mothers found out about their partner's escapades.
- An art collector in Munich received a shard from an antique vase as a gift from a friend. It just happened to be an exact match to a shard the art collector had bought in Athens decades before. Now, the girl and boy on the antique vase were once again harmoniously together as they had been 2,000 years ago in ancient Hellas.
- While allied forces were preparing to invade Normandy, the details of the attack were naturally kept top secret, including the code names for the different areas along the coastline: Omaha, Utah, Mulberry, Neptune and Overlord. Yet two weeks prior to the invasion, all of the code words appeared in a crossword puzzle in the London "Daily Telegraph". The shocked secret service subjected the author of the puzzle to intense interrogations. However, no explanation could be found. There was no doubt that the man was neither a traitor nor an agent.

Coincidence or Industrial Espionage?

The author himself was there in person when General Electric (GE) presented the first isotopically-pure, 12C diamond at an internal press conference in New York. The fact that this would be one of the most unusual events the author would ever attend was the result of a completely unusual coincidence, one that even today no one is quite sure whether or not it actually was a coincidence. But read the following report published in highTech magazine for yourself:

"Thomas R. Anthony, Chief Developer in the Corporate Research and Development Center at the US corporation General Electric Co. (GE) in Schenactady, New York, was at a loss for words. The scientist was reading an article an excited colleague had just given him in sheer disbelief. What was written in black and white seemed quite impossible: he was looking at an exact description of the first heat superconductor, a diamond of isotopically-pure carbon 12 – produced at General Electric's own labs. The best-selling US-American author Tom Clancy, had discovered the company's most spectacular and most secret research project and had incorporated it into a science fiction novel. The title: *The Cardinal of the Kremlin*.

Whether this was a particularly refined form of industrial espionage – as was the assumption in GE circles – or whether Inspector Chance had played a dirty trick on the company, has remained Tom Clancy's secret to this day, an author who has earned millions from his books – some even having been filmed, such as *The Hunt for Red October* – and yet who lives a relatively secluded life."

Of Lost Wedding Bands, Dentures and Steinway Grands

On October 24, 1996 the Berlin newspaper "Berliner Kurier" published the following report: "OSLO – A fairy-tale like story... Evelyn Nöstmo from the small town of Malvik in Norway had been looking for her golden wedding band for three years. And now she has finally found the valuable piece of jewelry - in the stomach of an elk that had been shot and quartered. Her wedding band had disappeared without a trace back in the winter of 1993. It had slipped from her finger when her car had broken down and she was forced to push it onto a small forest path. She had searched in vain."

About 15 kilometers from the site where her car had broken down, her husband Asbjörn shot a grand elk bull. The animal became Sunday dinner. As she was cleaning the cooking pot, she noticed a shiny object – her ring. "I almost fainted," reported the Norwegian glowing with joy. "The elk must have swallowed the ring as he was eating leaves and branches," supposed her husband, the mayor of the town. He assured: "I swear the story is true. Even if I am a politician."

On December 1, 1994 the German news reported: "Giant cod swallows dentures. – Cor Stoop still can't believe it. He has his dentures back. He had lost them on a boat trip on the North Sea on September 3rd. He became seasick when the wind increased to strength and was thus forced to the railing numerous times. He lost his dentures when he threw up over the side. They fell into the North Sea. But now they are back. A cod had unwillingly returned them. The 19-pound fish was fished out of the North Sea by the high sea fisherman, Hugo Slamat. As Slamat gutted the fish, he found the dentures in the fish's stomach. He informed the local radio station in Northern Holland. The station proceeded to broadcast the request that all denture wearers who had lost their artificial teeth in the North Sea should contact the station. Cor Stoop just happened to hear the broadcast, contacted the station, received the fisherman's address, drove to his home in Amsterdam, took one look at the dentures, and said, 'Those are mine!' When everyone refused to believe him he popped them into his mouth and they were a perfect fit."

I had booked travel abroad for my wife and myself in February of 1979. One evening, a few days before we were scheduled to depart, the phone rang. On the other end of the line was Sylvia B., a former classmate of my wife. They had not heard from one another since they graduated from high school seven years ago. Sylvia asked if she could visit us for a day or two. Under normal circumstances it would have been a pleasant encounter. However, we would already be gone on the dates Sylvia would be in town.

In talking to her, it turned out that Sylvia was about to take a vacation herself and had wanted to take advantage of our proximity to the airport in Frankfurt to pay us a short visit. Sylvia asked, "Where are you going on your trip?" "We're flying to Jamaica," my wife answered. "No way! That's where I'm headed!" Sylvia yelled into the receiver. The surprise was perfect when we heard that Sylvia had booked a hotel in Montego Bay, which was where we had planned to go as well. So, the reunion was held on Montego Bay Beach in Jamaica!

Another strange coincidence occurred ten years later in the Thai city of Phuket. Although we are not usually unfriendly to other tourists when we are on vacation, we do not go out of our way to initiate closer contact. This time, however, we met an American couple at our hotel complex situated right on the ocean and there was simply "chemistry" between us. As is common with Americans, we introduced ourselves by our first names. As the two bid us farewell before their departure the man gave me his business card. The last name on the card read "Steinway".

I couldn't believe my eyes. Our acquaintance was a descendent of Heinrich Engelhardt Steinweg, who changed his name to Henry E. Steinway after emigrating to the United States. There would not have been anything unusual about that, had Engelhardt Steinweg not originally come from our hometown of Seesen, where he had been an instrument maker. The first piano he ever built in 1835, he gave to his bride Juliane as a wedding present. The first grand piano he built in 1936 in an old laundry in Seesen that had been converted to a workshop. In 1851, he and his four sons emigrated to New York City where the family started out by working in various piano factories. In 1853, he started his own business. After anglicizing his name in 1854, the company was called Steinway & Sons. The business experienced an enormous upswing after winning first prize for its pianos at the New York Industrial Exhibit in 1855. Today Steinway is a citizen of honour in our hometown.

Another story that is somewhat strange is one I experienced in 1989 at a press conference in Madrid. At dinner, I got to talking about my studies with a somewhat younger colleague, Ulrich P.. It turned out that we had both studied in Göttingen with him starting his studies right after I had finished. I, for whatever reason, happened to mention that my wife and I had lived on the outskirts of town in a place called Geismar. "I lived in Geismar, too," responded Ulrich. "Do you happen to know the street with the funny name 'On the Paul'?" I asked him. "Of course, that's where I lived." Ulrich was flabbergasted. "What house did you live in?" "It was house number 2," I responded. "No way, it's even the same address." And indeed, Ulrich had moved into the same multi-family house in 1978, just months after we had moved out.

Coincidence or Telepathy?

Another unbelievable story took place in the spring of 2002. We spent the Pentecost holidays in a hotel on the ocean in Southern France. Our son Jeremy, who was nine at the time, quickly became friends with a boy about his age named Maximilian, whose parents lived in Starnberg near Munich. The two were almost inseparable and would have preferred their parents to dine together in the evenings as well. However, even though each found the other couple to be nice enough, neither was very interested in forming a closer friendship. Instead, we preferred to spend the evenings within family, driving through the area and enjoying each other's company over a cozy dinner in one of the typical towns of Provence. One evening we found ourselves in Gassin, a town about 7 miles from our hotel, when Maximilian and his parents suddenly appeared in the very same restaurant, even though there were numerous other options around. The boys were overjoyed. We, the parents, took it in stride.

The next day we drove to Grimaud for dinner, a town about 14 miles away. We had just found a parking space at the edge of the old part of town and gotten out of our car when Maximilian's parents appeared in the same spot, which caused his mother to reply, "The boys must be telepathic!"

As if that weren't enough: six months later – our vacation was long forgotten – we drove to Munich around Christmas time. We planned to have dinner at the Ratskeller restaurant, but Jeremy said how much he would love to go back to the Mövenpick café in the Stachus part of town. We thought about it for a moment and finally agreed. My wife and I were just handing our coats over to the coat check when Jeremy came running up excitedly: "Guess who's sitting in the restaurant!" It was Maximilian, of course…

Mr. Froböse, You Are under Arrest!

Sometimes coincidence manifests itself as a practical joke. I remember the year 1988. I was living with my wife in an apartment in Munich at the time, at Number 117 on Landsberger Street to be precise. There is a special reason I am mentioning the address at this point: before continuing I would like to point out that the rather uncommon name of Froböse is slightly more common in the southern part of Lower Saxony, but is very rare in southern Germany. There were exactly three entries in the Munich telephone book for the name at that time.

One morning as I was having breakfast with my wife, there was a violent knock on the door. I opened it and standing there were two men in their mid-thirties, who got right to the point. "Are you Mr. Froböse?" "Yes, what can I do for you?"

"Nothing, you are under arrest!" As I was fully unaware of having done anything wrong, I remained pretty calm. "That's a good one, you have any more?" However, the men were not very amused at all. As if upon command, they held their badges under my nose identifying themselves as members of Munich's crime squad. I was speechless.

Fortunately, one of the two policemen decided to clarify things: "We have an arrest warrant for Mr. Tobias Froböse, born in 1965 in Hildesheim." A little color returned to my face. "First of all, my name is Rolf and not Tobias, and second of all, I would love to have been born in 1965," I answered with a slight grin, being a member of the 1949 vintage. "That's odd," they both said in unison and asked for my ID, which they studied in detail.

"Hey, it looks like we have the wrong address," one of them said to his colleague. "By gum you're right, the guy we're looking for lives at Landsberger Street 171!" The officers excused themselves for having mixed up the numbers and given me a scare, and they disappeared as quickly as they had come.

By the way, I never found out what the other Froböse had done.

"Inspector Chance" in Science

Mankind's greatest achievements are often discovered by chance. One of the favorite anecdotes in this context is the story of how physicist Isaac Newton discovered the law of gravity. It is said that an apple fell on Newton's head in 1665. This occurrence made him painfully aware of the power of the earth's gravitational pull and inspired him to define the law of gravity. However, it seems that Newton himself made the anecdote up – the apple merely served to explain a problem he had been thinking about for years.

The discovery of penicillin, however, would hardly have been imaginable without the decisive influence of chance. In 1928, Doctor Alexander Fleming left his petri dishes filled with streptococci cultures lying open in his lab and headed off on vacation. Upon his return he noticed that mold had killed some of his bacterial cultures. The fungus was called *Penicillium notatum*. Yet, it took more than ten years before Fleming – with the help of scientists Howard Florey and Ernst Chain – was able to develop Penicillin from the experience. Fleming, by the way, was not the first discover the bacteria-killing properties of the mold: Doctors John Sanderson, Joseph Lister and William Roberts had been using a similar means to kill bacteria long before him. It was probably only chance that kept them from creating antibiotics from it, which would have made them the grand scientists.

Even physicist Wilhelm Conrad Röntgen would have failed to make his most significant discovery had he kept his lab more orderly. In November 1895, he was experimenting with gas discharge tubes. One day he noticed that some fluorescent crystals, that just happened to be lying nearby, lit up when he turned the tubes on. A yet unknown form of ray was emitted. Following this discovery, Röntgen left nothing else to chance. He systematically examined the rays, which he called "X rays" and which bear the very same name to this day. He radiated all possible sorts of matter with the rays and discovered that the "X rays" could be used to visualize the bones of the hand. In 1901, Röntgen received the first Nobel Prize for Physics for his work.

Another popular story is that of "Post-it" notes. The adhesive that did not really stick had already been invented – but nobody could figure out what it was good for. Chance came into the picture when an employee of the manufacturing company decided he wanted to mark the songs in his hymn book that he would be singing in church. A sticky label that could be easily removed was the solution. Is this anecdote of coincidence and adhesives really true? Whether it is or not, what is true is that the employee to whom the story actually happened, worked in the company's marketing department.

Coincidence alone does produce breakthroughs. It generally takes years of research before the coincidence occurs that plants the seed for the actual discovery. That is precisely what happened in the case of the French painter Louis Jacques Mandé Daguerre.

Daguerre painted large dioramas using the camera obscura that had come into fashion at the beginning of the 19th century. He projected images onto large canvases and then painted off of those images. In doing so, he began to develop the idea of a method that would perform this very tedious work for him: photography. Daguerre experimented for years until he figured out that silver plates that were vapor coated with iodine or bromine would retain the image from the camera obscura – however, it took hours of exposure time. Chance came into play one day in 1829 when he had to abort an exposure session because clouds had covered the sun. He put the plate back in his chemistry cupboard. However, when he reopened it later on, he discovered that the plate had been developed and it even showed a clearer and more detailed image than during his previous attempts. He did not discover the reason for this surprising development until later as he was removing the chemicals from his cupboard: a ball of quicksilver had formed in one of the joints of his cupboard. The steam from the quicksilver had developed the exposed plate. It still took Daguerre a few years to not only develop his photographs, but also to be able to retain them. The daguerreotype process, which was named after his discovery, became widely used during the mid-30s of the nineteenth century – until Henry Fox Talbot of England's discovery established itself: the negative-positive approach. This approach could be used to make as many copies as one wanted from a single exposure.

A Strange Dream Leads to a Fundamental Discovery

August Kekulé was born to a Hessian functionary in Darmstadt on September 7, 1829. After graduating from high school, he began to study architecture at the University of Giessen, pursuant to his parents' wishes, but the student soon became fascinated by the chemistry courses hosted by Justus von Liebig that he was secretly attending. He was so captivated, in fact, that he made his desires clear and began studying chemistry.

In 1856 Kekulé earned his professorship in Heidelberg. As there were no openings for him to chair a department there, he rented a house from a flour merchant. He used the largest room for lectures and the kitchen turned out to be a fine lab.

Despite the difficult circumstances, he managed a scientific breakthrough after two years: he discovered that carbon is tetravalent and that its atoms can form chains of themselves. This discovery is now part of the standard material taught in schools. However, in Kekulé's day, organic chemistry was still rather obscure. On the one hand, the number of organic connections was growing by the day and had led to an enormous gain in knowledge. However, practitioners tended to only characterize individual "trees", whereas theoreticians experienced limited success in their description of the "forest".

The fundamental theses of this newcomer were not accepted without criticism. Chemist Hermann Kolbe of Marburg in particular polemicized strongly against Kekulé, as a quote from the journal for practical chemistry showed: "The inability to organize his thoughts and to express said thoughts in an easily comprehensible manner, using clear wording, is only seconded by Kekulé's arrogance and his lack of gravity in the professional treatment of the matter. He proclaims the hypothesis he has invented and proposed of the so-called linking of atoms into molecules, referred to as the chain theory, to be of fundamental importance for the entire atomic theory and even to have become a law of nature…".

Chemists had an especially difficult time with a connection that was created the first time in 1825 from compressed coal gas and which Justus von Liebig referred to as "benzene". This material quickly revealed itself to be the progenitor of a new organic "family". The only problem: no one knew the formula for it, let alone its structure.

In 1865, Kekulé solved the benzene problem with his famous "hexagon rule". According to that rule, benzene had the simple formula of C_6H_6, and its structure corresponded to a standard hexagon with a CH group at each point. With a single move, Kekulé was suddenly able to characterize all of the "family members" that were known at the time and predict further reactions. Both the chemical and pharmaceutical industries profited considerably from this realization in the latter third of the 19th century. Thus, we have Kekulé to thank for the fact that chemists still have a fixed professional position in today's modern industrialized society. This is why he, unlike hardly any other scientist, was honored in Berlin in 1890 during the celebration of the 25th anniversary of the discovery of the formula for benzene.

Kekulé gave a remarkable speech during in which he mentioned that a dream had inspired him to establish the formula for benzene. What he said verbatim was, "During my stay in Gent in Belgium, I was living in an elegant bachelor's room on the main street. My office, however, faced a small side street and did not have any light during the day. For chemists, who spend their days in a laboratory, this was not a disadvantage. I sat there writing my teaching textbook, but it was not working very well as my spirit was focused on other matters. I turned my chair toward the fireplace and dozed off. Once again, the atoms swam in front of my eyes.

This time, smaller groups remained humbly in the background. My spiritual eye, sharpened by repeated history of a similar type, began to differentiate larger shapes of manifold structure. Long rows placed together in multiple thicknesses, everything in motion, winding and turning in a snake-like manner. But wait, what was that? One of the snakes grasped its own tail and the image swam before my eyes tauntingly. I awoke suddenly, as if struck by lightning. Yet again did I spend the rest of the night deciphering the consequences of this hypothesis."

This dream, which is often mentioned in psychology books and standard works on chemistry, was of great interest to two American chemists in the 1980s, who were involved in a scientific dispute over the truth behind Kekulé's quote. Professor John Wotiz, biochemists at the University of Illinois, relegated the dream story to the realm of legends. In his opinion, Kekulé's story even created a ruinous image of research chemists. It led Wotiz to comment in a journal: "A chemist experiments, collects data and then begins to formulate a structure. We need to get away from this benzene fairytale."

In his opinion, it was highly suspect that Kekulé's alleged dream was not mentioned in the presentation manuscript until 25 years after the discovery of the benzene formula. This led him to conclude that it was a joke. He claimed that the dream story was not in character with Kekulé's overall image, which had "always" been that of a "rational" personality in the field of research.

Wotiz, who at scientific conventions offered up for discussion his views on the Kekulé dream, found a vehement adversary in Professor Alan Rocke of the University of Michigan. As a chemistry historian, Rocke had taken an in-depth look at Kekulé and had published a 27-page article defending the dream theory. In contrast to Wotiz, Rocke saw no reason to doubt something that Kekulé himself had expressed. Rocke considered the allegations to be "naïve" that Kekulé waited to go public with his dream until 25 years later just to amuse his colleagues.

"Kekulé would have exposed himself to the risk of not being taken seriously. The same story, presented at a commemorative speech, had quite a different effect," he argued. At the same time, the scientific historian referred to a publication from the pen of Kekulé's son Stephan. In 1927, Stephan wrote that his father had shared the dream many times with friends and family. Rocke was further supported by American scientific historian Cyril Smith, who introduced the idea that many scientific discoveries had their origins in daydreams. He claimed this to be especially true of completely new theories that did not stem from the wide treasure chest of experience.

"It was not until later that a process we call logic was applied and the new idea that was grounded in scientific fact passed the practical tests," said Smith. August Kekulé's answer to the scientists' dispute would have sounded very similar. Taking another look at his published commemorative speech, we find what came directly after the dream story: "Let us learn to dream, gentlemen, then perhaps we will find the truth. But let us be wary of publishing our dreams, before they have been tested by our guarded understanding."

How a Chemists' Forgetfulness Helped Create a Super Substance

The best inventions are often the result of coincidence. The American Roy Plunkett is just one of many to have such an experience. The story follows that the 26-year young chemist had just fallen in love and was thinking about a date when he accidentally left a gas bottle filled with the fluorine compound Tetrafluoroethylene or TFE for short lying on the table in the lab instead of placing it in the refrigerator as planned. He had never forgotten before!

When Plunkett's lab assistant Jack Rebok opened the valve on the morning of April 6, 1938 to remove the compressed TFE, the bottle absolutely refused to emit any gas. But there was no way it was empty! As Plunkett could make neither rhyme nor reason of it, he suddenly decided to saw the steel cylinder open, where he found a white film. Subsequent testing proved a new plastic had been created – polytetrafluorethylene, later named Teflon. It was fire-proof and organic solvents did not cause it to swell even when exposed to the most aggressive acids - extremely uncommon behavior for that substance class to date!

However, the unique properties of the material would not be employed for many years – in the form of heat protection tiles, wire isolation and for astronaut's suits in aeronautics. That is why people like to refer to "Teflon" as a spin-off of the aeronautics industry. Although a nice story, it is nothing more than that. Teflon had already conquered frying pans as early as 1954 and not until after the "Sputnik Shock" in 1957 was it space's "turn".

Due to its excellent chemical and thermal composition, the material found a wide range of applications in the following decades, capturing numerous markets. The spectrum ranges from chemically stable piping for industrial facilities to medical injection tubes to microelectronics. There is no end to the "Teflon boom" in sight.

Can that Still Be Considered Coincidence or Is There a Hidden Higher Order Behind It?

"The word coincidence is blasphemy; nothing under the sun is coincidence."
(Gotthold Ephraim Lessing)

The Family Crest

This story involves my hobby. I have been interested in genealogy since the mid-90s. During my research, I came across a literary reference indicating that a family crest from the year 1667 was located in the sacristy of the St. Jacobi church in Lübeck. To acquire a photocopy, I would have had to travel 600 miles from my home in Upper Bavaria.

Fortunately, there is a regional mailing list for genealogists. So I entered my request in the mailing list for Schleswig-Holstein stating that I was looking for a genealogist located in the area who would acquire a photo for me in exchange for payment of all costs incurred.

I received a confirmation from another hobby genealogist just a day later. During our first phone call, I found out that said person did not actually live in Lübeck, but in a small town 30 miles from there. However, since I really wanted the photograph, I naturally offered to not only pay for the photographic materials (this was before the age of digital cameras), but also his expenses for the trip. He turned me down and said, and I quote, "I have the feeling that I need to do this for you." Extremely pleased, but also somewhat ashamed, I took the stranger up on his offer.

A few days later he called me up and was quite excited, "I have something unbelievable to tell you," he said with a shaky voice. "Good heavens!

Is something wrong with the crest?" I asked unknowingly. "You could certainly say so. I almost fell off the ladder as I was photographing it!"

He informed me that the Froböse crest in question was, along with a few other family crests, affixed to the wall of the sacristy, but at a height of 12 feet. In order to photograph it, he had had to borrow a ladder from the sexton. "And when I had reached the top rung, I was so excited, I had to steady myself before I could continue," he reported. "Because, right next to your crest hung that of my family!"

I am just as speechless today as I was fifteen years ago, when I think about this story. What inner sense was it that absolutely drove him to fulfill my request?

Zuckmayer's Curious Wallpaper

Carl Zuckmayer could hardly hope to encounter a second Carl Zuckmayer. He had to satisfy himself with a kind of "wallpaper of fate". Accordingly, Carl Zuckmayer found a hand-painted tapestry in an American Intellectual's home, which he had first seen at the Carl Mayr hotel near Salzburg when he was exiled to Austria many years before. "Many years after fleeing from occupied Austria," he wrote in *As if It Were a Piece of Me*, "I was once pulled away from the solitude of my Vermont farm and forest in America by friends to go meet an American author who had settled just a few hours' drive away in a town in old, colonial New England."

After an extensive tour of the house and after much urging by Zuckmayer, the author finally opened up a final unheated, and thus uninhabited, garden room, where on the wall the original tapestry from Salzburg was neatly hanging, "…as if Carl Mayr had just applied away the last touches of paint."

The Mysterious Double Exposure

A mother from the Black Forest shared the following unbelievable story at the beginning of the 20th century:

The woman had photographed her 4-year-old son, taken the film to Strasbourg to be developed, but was then unable to pick it up again as the First World War broke out just a few days after she had dropped off the film. Two years later she purchased a new film in Frankfurt to photograph her daughter who had been born in the meantime. After developing the film, she mysteriously found the pictures she had taken of her son two years before. The entire film turned out to be double exposures. The old film that had already been exposed must have somehow ended up back on the shelf in the shop where she had bought it.

It Doesn't Always Have to Be Plum Pudding

The next unbelievable story comes from the author Émile Deschamps, who lived in the 19th century. As a student in Orléans, he and the other children were always invited by Monsieur de Fontgibu to have plum pudding. Ten years later, Deschamps sees a piece of plum pudding in a Parisian restaurant and orders it in memory of his childhood event. However, it turns out that piece has already been reserved – for a certain Monsieur de Fontgibu. He offers to split the plum pudding with Deschamps.

Another ten years later, Deschamps accepts a lady friend's invitation to dinner. For dessert there is plum pudding. Just as it was about to be served, he and his friend joke that the only thing missing was de Fontgibu. At that exact moment the housemaid reports the arrival of a certain Monsieur de Fontgibu.

Deschamps thinks it a joke until an older gentleman enters the room. And who should it be, but Fontgibu, who had been invited to another party at the same address and who had accidentally gone through the wrong flat door.

The Dialog, Part 2: Some Doubt Arises

"Chance comes from Providence. And man must mould it to his own designs." (Friedrich von Schiller)

"Hmm, you seem to have a considerable collection here." Al places the materials aside.

"And what do you think of the collection as a whole?" Zach throws a questioning look at his friend.

"The stories are, without a doubt, becoming more and more exciting!"

"Just as I had promised."

"Yet, I am unable to identify a higher order or anything like it, coincidence is a fickle companion."

"There are a few cases where I am not so sure about that," countered Zach.

"And which are those?"

"I am referring to stories like the one with the crest, because not only does it involve coincidence, but intuition as well."

"Even that kind of gut feeling is completely normal. Sometimes I make decisions based on my gut feeling too and find out later that they were right."

Zach shakes his head: "No, I am referring to something else. When someone has the feeling they just have to do something, when they follow their inner voice in other words, and then something remarkable happens, I tend to see that as having a very different quality to it."

"And which would that be?"

"I am not quite ready to let the cat out of the bag just yet; the evening is still too young for that. But I will say this: according to Carl Gustav

Jung's psychology, intuition is one of four basic psychological functions that enable future developments, with all of its options and potential, to be perceived. It is usually perceived as instinctive understanding or emotional premonition."

Al leans back: "I can see that you plan on torturing me a bit today."

Zach gauges Al's response: "Admitted. I actually do plan on taking a closer look at this phenomenon and at C. G. Jung. But let's grab a bite to eat first, then we will attack the next round – you are hungry, aren't you?"

Al grinned widely: "You just happen to be right."

Why Psychologist and Psychiatrist C. G. Jung Was So Fascinated by Coincidence

Carl Gustav Jung, a student of Sigmund Freud, did not believe in coincidence. Instead, he considered the phenomenon of coincidence to hide a higher order, one that steered our lives. C. G. Jung and Sigmund Freud were once friends. Yet, Jung's repeated criticism of Freud's libido theories and understanding of the role of sexuality and the nature of the subconscious, drove a wedge between them. Even today Jung's interest in the phenomena such as synchronicity is still often misunderstood. However, those who consider his life's work to mysticism in scientific clothing, oversees the fact that he always took great pains to maintain the most objective perspective possible.

His principle of synchronicity describes how so-called coincidences can influence our lives. Within that process, synchronized events are linked to one another acausally – in other words, not by a chain of cause and effect. C. G. Jung understood synchronicity not merely as simultaneous events, but more as a similarity in meaning in which external events in a specific person's life correspond in a way that could not be explained causally. This differs from merecoincidental experiences, which simply show parallel events that are not actually connected.

Psychotherapist Robert H. Hopcke made the following comment on Jung's principle of synchronicity: "In my opinion, the phenomenon of synchronicity Jung has described – the simultaneous occurrence of two events that share similar meaning but are not causally connected – is still disputed and difficult to accept today because it forces us to shake off the unconscious tyranny of causal thought. Causal thought gives us the illusion that we have full control over our environment and reinforces our belief that we hold our fate in our own hands – an idea I hold to be very flattering."

Hopcke continues: "When we manage to see something as part of a whole, it begins to seem as if it was not we, our collective I, that has triggered a result, but as if there were a certain higher order cause, the Self written with a capital "S", the structure and coherence of which holds an exceptional plan for our life."

Researcher Alan Vaughan of Los Angeles continued the advancement of synchronicity research. In his book *Incredible Coincidence. The Baffling World of Synchronicity*, in which he shares 152 documented cases including their sources, he drafts a four-tiered classification of synchronistic events that range from trivial simultaneous events to in-depth experiences. He emphasizes the frequency of synchronistic events and considers them to be an indication of a fundamental principle of order within the universe. He believes that synchronicity is co-shaped by the human conscious, which he considers holographic. "Each element has knowledge of the whole. Just as each cell in our body has information on the entire body…so does each human have information on the entire universe," is his theory. There is also probably some truth to the common saying that two people are "on the same wave length".

The Discovery of the Periodic System of Elements – A Case of Synchronicity?

One famous example of a synchronous connection that is commonly cited comes from science: the discovery of the periodic system of the elements by both the German doctor and chemist Lothar Meyer, as well as by the

Russian chemist Dmitri Mendelejew in 1868 / 1869. They structured the chemical elements in increasing order according to their atomic mass, also they structured elements with similar properties above and below one another. This discovery, which was fundamental for chemistry, came about in a completely parallel manner, since neither scientist was aware of the other. Today, the periodic system serves mainly as an overview. Historically, it was very significant for predicting the discovery of new elements and their properties.

Another example comes directly from Carl Gustav Jung himself. In his collected works, he reported the occurrence of the following event (volume 8, page 497):

"During a decisive moment in her treatment, a young patient had a dream in which she was given a golden scarab beetle as a gift. I sat with my back to a closed window as she told me about the dream. I suddenly heard a noise behind me, as if someone were gently knocking on the window. I turned around and saw that an insect flying about outside kept knocking into the window. I opened the window and caught the insect in mid-flight. It was the closest analogy to a golden scarab beetle that could be found in our region, namely a scarab weevil, Centonia aurata, the mean old rose chafer. This chafer, in contrast to its habitual practices, obviously suddenly felt the need to be in a dark room at that very moment."

Another example of synchronicity comes from Graduate Psychologist Dr. Theodor Seifert:

"A couple with whom we are friends lost their 17-year-old son to sudden cardiac death three weeks before Christmas. In spite of their deep grief, their daughter, the sister of the dead boy, bought a Christmas tree shortly before Christmas Day.

As they were setting up the tree on Christmas Eve, they saw that one of the four branches at the top was now missing even though it had been there before. It had broken off. The sight of it renewed their sadness as the now three-branched crown of the tree reflected the family's three-person existence, the parents and the daughter – the son was no longer there, his existence had also been broken off."

Mrs. Elisabeth Hofbeck of Charlottenburg in Berlin shared the following story from her life with me:

A Dream Becomes Reality

"A few years ago, my daughter and I visited my girlfriend Val in Key West, Florida. One afternoon a few days after my arrival, I fell asleep in front of the television. My daughter was playing and my friend was at work. I dreamt that she had come home and had gone off shopping with my daughter. In my dream I then suddenly found myself with Val in a large portal. I saw a large, round aquarium that reached to the ceiling. I was so fascinated, that I went up to it and watched the fish from close up. Everything had been done true to nature, making it just like the ocean. I suddenly noticed three beautifully-formed stones and I said to Val: 'Just look at that. Nature is so beautiful in all of the different forms it has created.'

I then looked across the way and saw an open restaurant. It looked like it was no longer in business. Wooden chairs and wooden tables, parquet flooring, there was a bar straight ahead and a black piano to the left of it. I entered the room and looked around. Then I went behind the bar. At that moment I realized that Val was no longer there. Instead, a man carrying a full tote bag was coming and I suddenly understood that the restaurant was still closed and he was part of the staff. As I was embarrassed by the situation, I hid behind the piano. The employee began to wash the floor. Finally, he discovered me. He was furious, because he thought I was supposedly responsible for the mess.

The dream ended abruptly at that point, as Val had woken me up. I really struggled to get back to reality. My friend told me that she had come home earlier, but that I was sleeping so peacefully she had not wanted to wake me. I told her about my dream and how odd it was that I had dreamt my daughter was out shopping. I told her all of the details and she listened attentively.

After work the next day she offered to show me something. She drove us to a hotel and, at first, I thought nothing of it. But when we went into the reception area everything seemed very familiar, I just couldn't place it. I went up to the round aquarium, admired it, and said to Val: 'Look how beautiful these stones are.' Then she pointed to the open restaurant and said: 'What do you have to say about that?'

It also seemed very familiar to me, which I proceeded to tell her. It was not until then that she said: 'This is what you told me yesterday, this is what you were dreaming about.' I was completely perplexed and then I remembered why it seemed so familiar. Val told me she knew of the hotel and had wanted to surprise me, which is why she hadn't mentioned it the day before.

As the hotel was fairly new, it couldn't have been a memory from an earlier life. Had I left my body while I was sleeping? That is what Val thought, but that still seems incomprehensible to me. Although I have had a few dreams that turned out to be true, I can't remember ever having a dream like this one."

Beatles Forever

Mrs. Sonja Nagel of Bielefeld sent me a story related to her affinity for the Beatles. The author himself is an old Beatles fan. "I still have many of the band's records and even an autograph from the Beatles," she wrote me. Her e-mail continues, "I work at the university. One morning there was not a lot do to, so I sat down at my computer and researched what sort of things about the Beatles could be found on the Net. I was particularly interested in collectables, real signatures and collectors' corners. At that moment a student walked into the room to ask about the certificate for a course he had taken. I was very surprised as he was wearing a t-shirt with a picture of the four Beatles…"

Mrs. Nagel also contributed the following:

Death of a Newscaster

"There was once a newscaster named Hans-Joachim Friedrichs. Years ago when I was visiting my mother a picture in the BILD newspaper caught my eye: a fat headline reported that said newscaster had cancer! I was very sorry, especially as I found him very attractive.

At some point in the future I had a dream: I was facing an empty alley and it was pretty dark. I suddenly noticed a television in the middle of the street that was sitting on top of a small table. On the screen I saw Hans-Joachim Friedrichs speaking about something. However, I was unable to hear anything. As I was watching the man, I got the sudden feeling that he was dead. It was almost as if he had predicted his own death…

The next morning as I was drinking my coffee, I heard a report on the radio that newscaster Hans-Joachim Friedrichs had lost his battle with cancer the night before."

The Peculiar Affairs Had by Ann

In one of his works, American psychotherapist Robert H. Hopcke describes the story of a woman named Ann who had a series of affairs following her divorce. One of these relationships began during a vacation in Mexico. The man's name was Dan and he had recently separated himself inwardly from his wife. Even though the two lived about 100 miles apart, they continued their relationship after their vacation.

The passionate affair lasted a year. But then Dan realized he did belong with his wife to whom he had been married for many years. Ann tried to overcome her pain and to carry on with life as usual. She thought of Dan often, but resisted the urge to call him. A year went by. One day when one of Ann's good friends unknowingly proposed a trip to the coastal town where Dan lived, she hesitated at first, but then finally agreed.

About this Hopcke wrote: "It was one of those wonderful spring days that make you feel like you're in love, even when you aren't. And as could be expected, Ann thought she felt her lover's presence everywhere she went. But she did not manage to call him and also refrained from stopping by his house or his boat...Instead, she rambled through the city and shops with her friend, had lunch at a restaurant on the beach, and watched a beautiful sunset over the ocean before heading home.

As she arrived home and unlocked the door, the telephone began to ring. It was Dan on the other end of the line, whom she had not heard from in a year. 'Hello Ann, I was thinking about you today for some reason. So I thought I would call you up and see how you are doing."

The World Is Quite Literally Small

Although not a case of synchronicity, an extremely unusual coincidence did take place during a trip to England. I received this story from Professor Hans-Karl Müller-Buschbaum, my wife's cousin, who e-mailed it to me. This remarkable event involved his now deceased aunt Fanny Blaurock and her husband, who came from Wallendorf near the town of Lauscha in the Thuringian Forest. The following is a direct quote from his e-mail: "Karl and Fanny Blaurock were in England when they went into a small restaurant near London. Unfortunately, someone was already sitting at each table so that they had to ask whether they could join someone at one of their tables, even though this was quite contrary to their natural inclination (this is even more pronounced with the British than with the Germans). They started up a conversation with the British couple in which the Blaurocks told them they were only visiting England and that they lived in Dear Old Germany. The English couple then recounted that they had also once lived in Germany, 25 years ago in the Thuringian Forest in a small town named Wallendorf. There they had witnessed a wedding which they described in detail to the Blaurocks. It turned out to be the Blaurock's wedding!"

A Grandfather's Mysterious Inspiration

The author found a rather typical case of synchronicity in a report published by psychotherapist Dr. Elisabeth Mardorf of Bad Essen near Osnabrück: "The following chain of coincidences seemed so strange to the parties involved that a large report about it even appeared in the local newspaper: an exchange student from America had just arrived at his German host family's home two days before and had not yet called his parents back home. His grandfather was worried and tried to contact him over the Internet. Since the boy's host parents did not have an Internet connection, he looked up a random number in the local directory and ended up reaching a Mr. H. He asked the man to contact his grandson.

Mr. H. planned to do so later that day – it was a Saturday – and told his girlfriend, who managed the town pharmacy, about the story over lunch. He had hardly mentioned the name of the host family when the emergency doorbell rang: the customer who so urgently needed a medication was the American student's host father! When the pharmacist asked him whether it was for his American guest, he was extremely surprised. He did not even know the pharmacist. How could she possibly know about his American guest? Obvious in this story, is that the grandfather had been right to worry: his grandson was sick and he 'just happened' to pick the pharmacist's boyfriend as his Internet partner."

What Gave an IT Expert Goosebumps

The following story was told to me by Robert H. who works as an IT expert for a major communications company in Frankfurt. A few years ago he had suffered a health problem. Since he always felt run down, he went to his general practitioner who did a blood lab and found his white blood cell count to be extremely low. The GP said it was a condition called agranulocytosis. This was his first serious illness and he wanted to get to the bottom of it.

Robert had never heard of the condition before, but the diagnosis bothered him so much that even after returning to work the day after the tests, he kept thinking about it throughout the day. He tried to distract himself with his work, which he did manage to a certain extent. A meeting with a business partner also promised to provide the necessary diversion. Robert headed to Frankfurt-Niederrad around lunchtime, took the elevator up to the fifth floor of the glass office building and headed to his business partner's office, who had invited him to lunch. "Would you mind taking a seat for a moment?", a friendly secretary asked as she took his coat and offered him a seat. Before sitting down, his eyes briefly glanced at the secretary's computer screen. On the screen he saw a single word written in all caps: AGRANULOCYTOSIS. Robert H. assured me he had trouble keeping his composure!

"Preliminary Round" of a Class Reunion

One weekend late in the summer of 1999, I went to Seesen to attend my 30-year class reunion. During the six-hour train ride, I fell into a conversation with some other passengers. As it turned out, they were teachers who were just returning from a short trip to Munich. That was reason enough for me to tell them about my upcoming class reunion. "One of our colleagues was unable to join us because he has a class reunion," one of the teachers told me. "He would have been able to combine it with our plans, but he didn't want to stress himself out."

An inner voice told me that I should dig deeper. "Which school do you teach at?"

"We are from the junior high school in Langelsheim." "Oh, that is not far away from my hometown of Seesen," I replied. "What's the name of your colleague who is going to the class reunion?" The person sitting across from me looked at me questioningly: "He's a very nice colleague by the name of Jürgen B., do you happen to know him?" "And boy do I," I responded. "We graduated from high school together back in 1969 in Seesen and I really look forward to seeing him again after 30 years."

Greetings from Venus

A strange coincidence happened to the author himself when he was on a business trip. The reason for the trip was a technical press colloquium hosted by Alstom Deutschland GmbH and being held on November 29, 2007. A podium discussion was held toward the end of the event on the topic of "Chance for an Objective Climate Debate".

First came a presentation by an expert, who presented a series of climate scenarios he had created using computer simulation. It all became too much for me when, during the subsequent discussion, a participant voiced his opinion that he did not believe in the greenhouse effect of carbon dioxide at all, citing analyses performed by renowned experts. Being a chemist, I pointed out that carbon dioxide has the strongest absorption potential precisely in the infrared zone and that any allegations to the contrary were unsound.

This was met by numerous objections, including they were "only" computer simulations that could not simply be projected to larger planets. To which I responded that Venus is a planet that is about the same size as the Earth and that due to its proximity to the sun it must have an average surface temperature of 140 °F if the atmosphere was anything like the Earth's. In reality, the surface temperature of Venus is a hellish 840 °F, because the planet's atmosphere consists of 95% carbon dioxide. In other words: the surface temperature is about 700 °F higher than one would assume based on its distance from the sun. Thus, Venus, which is 67 million miles away from the sun, is hotter than the Mercury, which is only 36 million miles from the sun. I mentioned that was something we might want to look into! I had brought the issue up during the discussion out of a spontaneous sense of need.

Purely intuitively, I publicly speculated that Venus may have originally had oceans in which the carbon dioxide, produced by the volcanic emissions caused by the high temperature of the water, was prevented from being released, let alone harden in the form of calciumcarbonate. I continued by saying that the increasing level of carbon dioxide finally caused the temperature at the surface of the planet to exceed the boiling point of water, which sealed Venus' fate, resulting in death by overheating. At the end of the event, I fell into a discussion about the topic with a few colleagues.

Following the conference, I drove directly to the airport in Cologne/Bonn. As I had a little time before my flight was scheduled to leave, I went into the Lufthansa Frequent Travelers' Lounge. Standing in front of the magazine rack, I picked up a copy of the daily newspaper "Die Welt", flipped through the political and economic sections, and then stopped at the second-to-last page of the science section. I couldn't believe my eyes. The latest report from the ESA, which I will not go into in greater detail than to share the headline, read:

"There Used to Be Oceans on Venus As Well / Greenhouse Effect Caused Temperature to Surge Up to 840 Degrees"

Paris – Even Venus had oceans just as Earth at one point in time. Scientists at the European Space Agency (ESA) came to this conclusion after evaluating data delivered by the European space probe "Venus Express". Yet, in contrast to our blue planet, a hostile atmosphere developed on the morning and evening star that consisted of almost 95% of carbon dioxide (CO_2) and where temperatures exceeded 840 degrees. "The water that previously existed on Venus evaporated. There are still traces of it in its atmosphere and we can still observe the process," stated ESA scientist Hakan Svedhem.

That the water on Venus evaporated, but not the water on the Earth is probably related to the planet's closer proximity to the sun, at a distance of 67 million miles from the sun. The Earth, on the other hand, is 93 million miles away. "The atmosphere was slightly hotter, which caused it to create more steam, which has a much stronger greenhouse effect than CO_2," reported Svedhem in the scientific journal "Nature". This caused the temperature to rise even more.

According to scientists, the atmosphere on Venus was initially composed of steam and carbon dioxide, like on Earth and Mars. On Earth, however, a large part of the CO_2 ended up binding, in the form of calciumcarbonate deposits on the Earth's crust, in the oceans and in living organisms. On Venus, however, the CO_2 content was so high after the water had evaporated that the greenhouse effect was further accentuated.

Yet, after a total of 30 missions to Venus since 1962, the planet still has not revealed all of its secrets. A thick coat of sulfuric acid-filled clouds still prevents us from capturing images of the planet's surface structure. A special light spectrum called the "Venus Express" was able to analyze the composition of the atmosphere. This allowed ESA researchers to determine that the wind in the higher levels of Venus' atmosphere blew at a speed three times that of a hurricane and that the temperature differences between day and night fluctuated up to 70 degrees Fahrenheit.

Europe's first mission to Venus was sent out in November 2005. The probe reached Venus' orbit in April 2006 and has been sending back data about the planet on a regular basis ever since. The mission is scheduled to end in May 2009.

An Odd Coincidence on Christ the King Sunday

The following is yet another story the author himself experienced: On November 24, 2007, a Saturday, my wife suggested we attend the service at the Wasserburg Church of Christ the following day. I thought it was a good idea since it was Christ the King Sunday.

When we arrived at the church that Sunday around 9:50 a.m., it was still pretty empty. After my wife and I took our seats in the second row, I began to think about my father who had passed away two years before and who had been laid to rest at the cemetery in our hometown of Seesen. Then I thought of my mother who was attending a service 400 miles away at the St. Andreas Church in Seesen.

Even before prayers began, we were greeted warmly by the Reverend Mrs. W. from Schnaitsee. She held a warm-hearted sermon and finished by reading the names of the members of the community who had passed away that year of 2007, as was the tradition for Christ the King Sunday. The age and hometown of the deceased were also mentioned. Since we live in a very down-to-earth area, most of the deceased were from Wasserburg or nearby communities. In the middle of the reading, my wife looked at me and said: "Did you just hear Seesen, too?" I shook my head, "I think she said Edling." And that was the end of our conversation.

As we headed toward the exit after the service with the church bells ringing, we realized that the church was almost full. Sitting at the very front, we hadn't even noticed. We were leaving the parking lot when two of my wife's good friends, Karin J. and Luise P., headed straight toward our car. I opened the window and my wife and I greeted them heartily. "Did you hear there was a woman from Seesen among the deceased?" Karin J., who was familiar with our hometown 400 miles away through her acquaintanceship with my wife, looked at us inquisitively. Luise P. confirmed it. My wife had heard the name of the town correctly after all!

As I was talking to my mother on the phone the next day, I told her about the odd coincidence. She asked whether we happened to remember the name, but we didn't. Thus we decided to call Mrs. W.

I have to admit that I was a little hesitant to contact the reverend about a matter that could be interpreted as mere "curiosity"; however, since our son Jeremy was friends with her son Marcel, I was fortunately able to overcome my hesitation.

When I explained my concern to Mrs. W. on the phone, she was very friendly and was able to tell me the name of the deceased woman from Seesen, a Mrs. H.. She also gave me the woman's son's telephone number, who, like the reverend, lived in the community of Schnaitsee close by. I thought about it for a moment and picked up the receiver.

I told Mr. H. the reason for my call and as "old Seeseners" we quickly fell into conversation and talked about the past. We had both gone to the same high school, but had not met as youth since we were seven years apart in age. In surprise, we realized that in 1970 we had both lived on Lautenthaler Street in Seesen and our parents had been friends with the family who had lived next door, Family Fricke.

When I called my mother shortly thereafter to tell her the name of the deceased, she was very sad. She told me she had not heard from Mrs. H. in the recent past. The last time my mother had heard from her had been five years before around Christmas time, when they had run into one another on the same train. My father was still alive at the time and my parents were traveling to Reitmehring via Rosenheim to visit us over the holidays. Mrs. H. continued on to the next stop, Bad Endorf, where her son was waiting for her.

Pseudonyms and Reality

The following story comes from a colleague. He asked me not to mention his name or his pseudonym, since he feared it might get him into trouble. The point is that this colleague, an economic journalist, had to assume a pseudonym against his will at the beginning of the '90s as he was working as a freelance journalist writing a series of "specials" for a large business publication. As the same name could not be used for all of the contributions, the chief editor urged him to assume a pseudonym. I will simply refer to him as "Kai Petzold", as his real pseudonym can still be tracked in business databases. He assured me the name had come to him out of the blue when the matter was proposed to him.

About ten years later when "Kai" had been writing for other publications for quite some time and no longer used his pseudonym; he was reminded of his past in a very odd manner. It was a few weeks before his 50th birthday.

In honor of the event, Kai had invited numerous guests – relatives, old and new friends, as well as colleagues. In preparation for his speech, Kai entered his birthday on Google, but left out the year. His search produced the names of a few famous personalities, who had been born the same day he had, the 2nd of August. He decided to add the year after all. This search produced a few news articles from the day he had been born from which he was able to collect further material for his speech.

One final document that he didn't quite know what to do with was a forum hosted by the city of Munich where hobby chefs could exchange recipes. Just for fun, he began to scroll through the entries in search of his date of birth. When "Kai" finally found it, he was flabbergasted. Among the entries was one by a man who lived just a few kilometers from his home town and who had come into the world the same day he had, a so-called "astro twin".

That in of it itself would not have been that unusual, but the man's name was Kai Petzold! My colleague still wonders why that specific name had spontaneously come to his mind while he was searching for a suitable synonym.

A Church Choir's Guardian Angel

Another incredible story was published in a 1950 edition of the renowned Life magazine. The story was about a church choir in the small town of Beatrice, Nebraska that was scheduled to assemble on March 1st at 7:20 p.m. for a rehearsal, yet every single one of the 15 members of the choir arrived late that evening for different reasons. The reverend's family was doing their washing that day and did not get done in time, one member's car refused to start, another was listening to the radio and lost track of time, one singer had to finish her homework, and one mother had trouble waking up her daughter, who had fallen asleep, causing them both to be late.

That each of them accidentally (?) missed the beginning of the rehearsal turned out to be quite a piece of luck, as an unidentified defect in the heating system caused a massive explosion at the church at 7:25 p.m. In interviews with Life magazine, the members of the choir questioned whether the synchronicity was not actually an act of God?

First Physical Interpretations of the Phenomenon

Psychologist and psychiatrist Carl Gustav Jung maintained a lively correspondence with physicist and Nobel Prize winner Wolfgang Pauli, which began in 1947 and in which the two searched for physical interpretations of synchronicity. Together they proposed not only including the dimensions of space and time, causality and the law of the conservation of energy in the formula used to explain nature, as had been common practice to date, but to include synchronicity as well. However, that would mean considering synchronicity not only as a rare occurrence in nature, but as a completely normally occurring phenomenon.

Wolfgang Pauli, who was born in 1900 in Vienna, was one of the most talented scientists to have ever lived. After graduating from high school in 1918, he began working on his dissertation in 1921 at the age of 21 under Arnold Sommerfeld in Munich. After completing his dissertation, he acquired his professorship in 1924 in Hamburg. In the past he had heard presentations on atomic physics held by Niels Bohr in Göttingen. Pauli formulated the exclusion principle in 1925 for which he received the Nobel Prize for physics.

The Pauli principle, or exclusion principle, states that two electrons with the same momentum and the same spin cannot occupy the same space. According to the Pauli principle, electrons can thus not occupy the same quantum state.

Although the principle sounds quite abstract, it does have far-reaching consequences. If we examine the principle under a scientific magnifying glass, we find that according to the Pauli principle, the electrons of an atom orbit around the nucleus on different orbital paths. Without the Pauli principle, one would assume that instead, atoms collapsed, because all of the electrons would jump to the innermost path as it had the lowest energy. From a popular science point of view, the Pauli principle is the reason you, dear reader, do not fall through your chair to the center of Earth, because it states that atoms are inpenetrable.

There is a nice anecdote about Wolfgang Pauli that I would like to mention at this point as coincidence plays a dominant role in the story. The story is about the strange coincidence that all experimental equipment tended to fail or even break in Wolfgang Pauli's presence. Pauli himself even believed in the objective existence of the effect, as were a few of his colleagues. Experimental physicist Otto Stern, for example, banned him from his lab. This so-called "Pauli effect" is jokingly referred to as the "second Paulian exclusion principle" in allusion to Pauli's famous exclusion principle. The Pauli *effect* is defined as: "It is impossible for a Wolfgang Pauli and a functioning device to occupy the same room."

An event that became quite famous occurred in James Franck's lab in Göttingen in which a valuable and sensitive part of a device broke when Pauli was not present. Franck told his colleague, who was living in Zurich, about the event and jokingly said that at least Pauli could not be blamed in that case. Pauli, however, replied that at the time in question, he was on a train to Copenhagen which had, at that moment, made a short stop in Göttingen.

How the Soul Became Taboo No Longer

Surprisingly enough, the intense correspondence between Jung and Pauli over half a century was suddenly no longer a topic of discussion. It seemed that the idea that the state of the soul and the uninhabited world could be connected and their effect on each other was considered too foolish of an assumption to even be discussed among elite scientists. Not until recently have the most recognized researchers been apparently willing to permanently remove the topic from the list of taboos. Among these researchers is the now retired Prof. Hans-Peter Dürr, one of the most famous quantum physicists of our time. Dürr was born in 1929 in Stuttgart and earned his doctorate in 1956 under Edward Teller, the father of the hydrogen bomb. Together with Werner Heisenberg, a founder of quantum physics, he did research from 1958 to 1976. Dürr was head of the Max Planck Institute for Physics in Munich from 1971 to 1997. I will get back to him, as well as other researchers who have argued theses that have led them to being called revolutionary, somewhat later on.

The Dialog, Part 3:
Coincidence Does Not Always Equal Coincidence

"If you keep up with all these examples I will wake up tomorrow morning a converted esoteric." Al lets out a deep sigh.

"And why ever not? With esoterics like Wolfgang Pauli you would be in the best of company," Zach responds spontaneously.

Their jabs at one another made them slightly giddy. "I was just thinking about my doctoral adviser. He was a huge fan of Pauli," grinned Al. "Had I ever referred to good old Pauli as an esoteric during my graduate or doctoral exams, I probably would have failed the test."

"And for good reason!" Zach leans back relaxed.

"Well – I think your examples are pretty crass." Al looks at him with a scowl.

"I really can't do anything about that. The sources are completely reliable! By the way, coincidence does not always equal coincidence. I think these examples make that very clear."

"Well, well – so coincidence does not always equal coincidence. What is that supposed to mean?"

"I want to show that the phenomenon of coincidence has two very different faces. The first, first-order coincidence shows the trivial side, a reality that can be explained by mathematical rules, but one that we allow to fool us time and again. We also try, in vain, to find a deeper meaning to them. The second face is that of a higher-order coincidence and is not actually a coincidence at all. It is based on a series of connections that science is just beginning to identify and research."

Al refills their glasses. "Good, then let us follow in Jung's and Pauli's footsteps and try to examine the matter from a scientific point of view."

"Exactly. I find an approach that combines all of the different aspects, such as strange coincidences, synchronicities or phenomena that fall into the field of parapsychology to be extremely exciting. The exchange between Pauli and Jung would have been quite impossible without that kind of interdisciplinary mindset."

Al gestures with his hand showing his skepticism. "Still, I think entertaining mind games like that border on science fiction."

Zach does not let up. "My dear Al, had the human spirit limited itself to traditional thought throughout history, we would still be living without electricity or heat. The most important scientific discoveries were made precisely because there were avant-gardists who dared depart from the realm of tradition to discover new terrain."

Spooky Physics:
A Short Excursion into Quantum Mechanics

I mentioned quantum physics earlier on, but allow me to comment on it at this point. Quantum physics is a topic that is so extensive I could easily write an entire book about it. Of course, many others have already undertaken that "job". Such works allow you to gain deeper insight into the matter. In this book, however, I will limit myself to the gist of it, to what you need to know in order to better understand the phenomena of coincidence I describe in the following chapters.

What is certain is that many quantum mechanical operations appear to be perplexing and hard to imagine, as they seem to contradict our everyday experiences. Even with this in mind, it is of little comfort that like Einstein's theory of relativity, quantum mechanics is one of the most revolutionary discoveries of the 20th century. For the purpose of this book, let us simply accept this fact for what it is.

The duality of light and particles is one of the core elements of quantum mechanics. Just as light not only has a wave-like nature, but also the properties of particles, typical particles – including electrons, nuclear atoms and even entire atoms – also have wave-like properties. The objects of quantum mechanics, however, are neither waves nor particles. Instead, they simply share the properties of both. Thus, in quantum mechanics, one generally speaks of 'states'. Many quantum mechanical states can be displayed using wave functions, although the wave functions describe both the wave-like and particle character of the quantum object.

In other words: a quantum does not behave like a tiny little pool ball. Instead, quantum theory describes the physical state of a particle using an abstract state function. There is no way to prove this in and of itself.

The square of its absolute value determines how the probability, physical size and the location or momentum of the quanta can be measured.

This is where quantum physics begins to differ from traditional physics and the "world as we experience it". In other words, the world we are familiar with. As a result, traditional physics allows the location and momentum of a specific particle to be determined at any level of precision without greater ado. Quantum mechanics, however, does not allow that type of determination. Nobel Prize winner Werner Karl Heisenberg was the first to formulate this fundamental "ban" back in 1927. The Heisenberg uncertainty principle states that no two measurement variables of a particle (such as both the location and the momentum) can be accurately measured at the same time. The uncertainty is not the result of an inefficient measurement process. Instead, it is more a matter of principle.

Let me use a short gedankenexperiment, or thought experiment, to clarify: an experimenter places a particle under the microscope in the hope of determining the particle's exact location. During this process, he continuously reduces the wave length of the light to increase the resolution. At least one photon must pass through the microscope and fall on the particle in order to be able to "see" it. This photon, in turn, has momentum p, which increases as the wave length decreases. The "measurement" of the particle's location itself triggers an impulse in the particle; the particle is basically being given a slight "push". This, in turn, prevents the exact location of that particle from being determined with any degree of accuracy.

Almost Esoteric: Strange Action at a Distance

Another quirk of quantum systems is the entanglement principle. It states that when two quantum systems interact, they then have to be viewed as a single system. The "entanglement" still applies even if their interaction lies far in the past and the two partial systems are now separated by great distances. The results of this effect are a reminder of supernatural phenomena, as another thought experiment shows.

In this case, an experimenter located anywhere in the world performs a measurement involving particle A. Particle A, however, is entangled with particle B. As a result, the latter is also simultaneously influenced by the measurement. It does not matter what distance separates particle A and particle B, whether 10 miles, 1,000 miles or even light years. And, as mentioned, they are influenced simultaneously – not at the speed of light, but infinitely fast!

The phenomena involving entanglement, in which a measurement taken at one location influences the measurement results taken at another location – no matter the distance, was actually one of the reasons Einstein rejected quantum mechanics.

Summary: the intricate, fundamental assumptions on which quantum physics are based are so different from the "world" with which we are familiar that we are forced to separate ourselves from our past experiences in order to understand them. Although that is easier said than done, there is really no way around it. In other words: we have to accept quantum mechanics just as they are. The same holds true for entanglement: the properties of two particles remain interdependent, even if they are at completely different locations. French physicist Alain Aspect proved this effect back in 1982. More recent experiments were performed by Austrian quantum physicist Professor Anton Zeilinger.

In principle, quantum theory allows communication between particles that is faster than the speed of light. Critics, who believe that assumption contradicts Einstein's theory of relativity, thus counter that the correlation explained by entanglement already existed and did not need to be transferred from one measurement location to the other. I do not wish to get into this modern discussion at this point, as I find it to be too "party political". However, I do find the phenomena that accompany entanglement to be extremely exciting and I will come back to that later on. What is more, the field is currently experiencing quite a lot of movement and we can look forward to seeing results in the near future.

No less exciting is the question of whether it will ever be possible to set a complex macroscopic system into a quantum state. I would like to share a chapter from one of my books that pertains to the technology of the future in response to that very question.

Will We Someday Be Able to Travel by Teleportation?

"Space, the final frontier. These are the voyages of the starship Enterprise. Her five-year mission: to explore strange new worlds, to seek out new life and new civilizations, to boldly go where no man has gone before…". Albert Einstein should not be on board when the captain engages the spaceship, travelling across the galaxy at "warp speed", because he would pull the emergency brake. That is because the laws of his theory of relativity, the speed of light at just around 300 000 kilometers a second, set relatively tight limits for interstellar travel and communication.

Physicists did not begin to question Einstein's dogma until the end of the 1990s. While experimenting with light particles, Nicolas Gesin and his colleagues from the University of Gent found, for example, that the particles were able to exchange information at infinite speed across a distance of multiple miles. The experiments performed by the team from Gent also tested the quantum theory, parts of which directly contradict Einstein's theory of relativity. Their tests showed that, in principle, quantum theory allowed communication between particles at speeds that exceeded the speed of light.

To break through the light barrier, the researchers transformed the city of Gent and surrounding areas into one giant lab. At the "headquarters" in the city, they directed a laser beam through a crystal composed of potassium, niobium and oxygen. The beam split the photons in two, each with about half of the energy. The researchers then directed the photons, which were now infrared photons, using standard fiber optic cables, sending them in opposite directions. One of them was sent to the nearby town of Bellevue, while the other was sent to Bernex, a town 11 kilometers away. The particles followed the known rules of physics until shortly before reaching their goal.

But then something astonishing happened. The cables in both cities split forming a Y right before reaching the measurement sensors. Photons in both Bellevue and Bernex had to "choose" between a shorter branch or one that was 20-centimeters longer. In some mysterious way, the split particles "knew" what their far away other half was doing, because they generally tended to choose the same branch. The researchers concluded that this faster-than-light communication between the particles worked not only for earthly distances, but also across light years. In other words: a lost photon, roaming around at the ends of the universe is generally able to spontaneously trigger a quantum effect on Earth!

Surprisingly, Albert Einstein himself had come across this strange phenomenon of entanglement, but he chose not to pursue it, instead referring to it as a "spooky action at a distance". But Einstein was mistaken, because spooky seems to work as one sensational experiment performed at the University of Innsbruck in June 2004 impressively proved. Although Professor Hans Briegel's team of scientists at the Institute for Theoretical Physics did not manage to create aggregates for spaceships that could move faster than the speed of light, they did manage a significant milestone that was no less seminal to the future of technology.

This discovery was nothing other than beaming itself. In real-time, in other words without any loss of time, the scientists transferred the quantum state of an atom to another atom ten micrometers away. This second atom then showed the exact same properties as the first. The experiment showed that information can be transferred instantaneously using matter – even if only across an extremely short distance. The first proof that teleportation was possible was found by Austrian researcher Professor Anton Zeilinger in Innsbruck in 1997, first using massless light particles: photons. Zeilinger, who now works at the University of Vienna, is one of the most distinguished "minds" of modern quantum physics and is now being "treated" like a "hot candidate" for the Nobel Prize for physics.

Captured calcium atoms, chilled to temperatures close to the absolute zero point (-273 °C) form the heart of a teleportation experiment being performed at the University of Innsbruck. Scientists use lasers to control the exact inner state of the atoms – their quantum states – and then entangle two atoms. One of the two atoms is then entangled with a third atom. This third atom then shows the properties, i.e. the quantum states that were supposed to be transferred. A measurement of the initial pair alone and a series of laser pulses sufficed to transfer the state of the atom to be teleported to the remaining atom.

The question is how this can be transferred to the dream of fast and comfortable travel using beam technology. Will future generations be able to slip into the role of Captain Kirk and Mister Spock and "dematerialize" and "rematerialize" at will? On the one hand, it remains completely unclear whether it will ever be possible to set a complex macroscopic system into a quantum state. Even if it did become possible, it would raise another question. "The information on the quantum states of a human alone that would have to be transferred in order to beam someone, would fill a pile of CDs 1 000 light-years long," Zeilinger tempers premature optimism.

Financial investors seem to see it similarly: after all, the reports of success coming out of Innsbruck have not yet caused airline stock to crash. But who today can know what the world will look like in the 23rd century? One thing is fairly certain: CDs will probably be a relic of the Stone Age of information era by then. And the quantum computers available in future, with all of their advancements, can probably solve the problem of storing the quantum states of macroscopic objects as well.

From Schrödinger's Cat to the Many Worlds Theory and Quantum Immortality

The transfer of quantum mechanics to macroscopic systems creates paradoxes that need to be discussed at greater length in science. The most popular example by far comes from the Austrian physicist Erwin Schrödinger. He proposed a thought experiment, which I quote here verbatim, "One can even construct quite burlesque cases. A cat is shut up in a steel chamber, together with the following diabolical apparatus (which one must keep out of the direct clutches of the cat): in a Geiger tube there is a mass of radioactive substance, so little that in the course of an hour *perhaps* one atom of it disintegrates, but also with equal probability not even one; if it does happen, the counter responds and through a relay activates a hammer that shatters a little flask of prussic acid. If one has left this entire system to itself for an hour, then one will say to himself that the cat is still living if in that time no atom has disintegrated. The first atomic disintegration would have poisoned it. The -function of the entire system would express this situation by having the living and the dead cat mixed or smeared out (pardon the expression) in equal parts."

Fortunately, no one has had the idea of actually performing this experiment. Not much would come from it either, because as the reader has probably correctly guessed by this point, you can simply open the chamber to find out whether the cat is still alive or not. Quantum mechanics would immediately argue that opening it – the observation in other words – would represent an intrusion that would produce an irreversible collapse of the wave function. In other words: the spooky "dead **and** simultaneous living" state of the cat would, at the end of the experiment, result in a dead **or** living cat.

Erwin Schrödinger uses the paradox of a dead and simultaneously living cat to emphasize the insanity of quantum mechanics, which enraged a few of his colleagues. "When I hear about Schrödinger's cat, I reach for my gun," is the way British astrophysicist Stephen Hawking once put it.

Just as crazy is the "many worlds interpretation" of quantum mechanics, which can basically be traced back to the American physicist Hugh Everett. Like other interpretations, the many worlds interpretation also attempts to attach a physical reality to the mathematical formalism of quantum mechanics. What motivated Everett was the aspect of the wave function collapsing, an aspect still accepted by most scientists today just as it was back then.

According to this interpretation of quantum mechanics, a system can progress over time in one of two ways: continuously or discontinuously. As soon as a quantum-physical state is measured, it collapses into a so-called independent state. The collapse is abrupt and immediate. Moreover, the disruption of the system caused by the measurement can have any degree of strength or weakness.

The many worlds interpretation also states that the Schrödinger equation describing the system is always unlimited and is unlimited everywhere. An observation or measurement of an object by an observer can be described by applying the equation to the overall system, i.e. the system composed of the observer and the observed. The result is that each observation causes the wave function to disintegrate into numerous "branches" or "worlds" that do not interact. In turn, the fact that an uninterrupted number of observation-like processes are taking place would lead to the conclusion that there must be an enormous number of simultaneously-existent worlds.

In one thought experiment referred to as quantum suicide, Canadian scientist Professor Hans Peter Moravec performed an experiment in the field of artificial intelligence in 1987 using the many word interpretation of quantum mechanics.

The native Austrian describes his thought experiment as follows: "The experiment is similar to that of Schrödinger's cat. A scientist sits facing a gun that is fired when a special radioactive atom decays. In this case, the scientist dies. According to the many world interpretation, the gun is fired at different times in the various parallel universes so that the probability of the scientist surviving the experiment because the atom does not decay is greater than the probability of his death. Even if the conditions are tightened (to an atomic bomb, for example, that explodes next to the test person), the theory still provides for a chance of survival. Looking at the system as a whole, the scientist does not die during the experiment because the probability of his survival is never equal to zero and thus he always survives in some universe."

From that perspective, the scientist is immortal, which is why the "experiment" is sometimes also referred to as quantum immortality. And that, as I mentioned before, is not esoteric, but one of the spearheads of modern physics.

Before I allow Al and Zach to continue their discussion, I would like to take a look at three other examples of stories of coincidence. Like quantum physics, these also remind one of something spooky. As a natural scientist, I am naturally fully aware that I am "going pretty far out on a limb" here. That is why I exclusively used serious sources which, to the best of my knowledge, I consider trustworthy in selecting these events.

Strange Encounters in Student Housing

After completing his degree in chemistry, the author began his scientific career at the Gmelin Institute in Frankfurt, which is part of the Max Planck Society. His scientific team included a few colleagues from the USA. One of them, Dr. James G. of San Francisco, was a very nice and friendly colleague with whom my wife and I enjoyed spending our time now and then. One night, James shared a story with us that he had rarely spoken of in the past.

"I studied chemistry not only in the USA, but also for a few semesters in London. When I came to England, the student housing was full, so I added my name to a waiting list. A short time later, I received the joyous news that a room had become available. Shortly after I had moved in, I awoke one night and in the twilight was able to see a young man with curly, black hair. I was terrified and told the alleged neighbor that he had the wrong room. He simply cried and looked at me with great sadness in his eyes.

"When I turned on the light, the apparition had disappeared. Since I was one hundred percent sure it had not been a dream, I told the housemaster about the strange encounter the next morning. I gave her a detailed description of the young man. She suddenly paled. Then she looked through the archives and showed me a photo. I immediately recognized the young man who had visited me in my room the evening before. When I asked her who he was, she replied with a quivering voice that it was the previous renter. She then added that my room had become available because he had taken his life shortly before."

The author would never have recorded the story had "James" not been an absolutely trustworthy person. Even the following story, which may also seem like a ghost story, comes from a trustworthy source.

Tragic Death of a Classmate

Gabriela Anna Kerer from Prutting (a town in the region of Rosenheim (Upper Bavaria) reminisced:

"I had a classmate in grade school that lived one kilometer from my house. When I changed schools, I no longer had any real contact with him until later when he occasionally stopped by the restaurant I was managing at the time. I lost all contact when I married and gave up the restaurant.

"Unfortunately, Markus was killed sometime later in a car accident. As I was sitting in the church during his funeral, I kept looking up at the Virgin Mary hanging behind the altar. I suddenly saw a driver driving around a corner. Behind the curve there was a car that was parked across the lane. I heard, 'Oh shit!' Suddenly, I was surrounded by darkness and it felt as if I had been packed in cotton. I felt light. After the moment had passed, I was certain I had just witnessed the last seconds of Markus' life.

After church I told the story to a good acquaintance. He said it had not been a trick of the mind at all as the mother of the deceased had seen the same."

Mrs. Sandra Menzel, in turn, told me about the following case:

The Grandma from the Afterworld

"My daughter was born in August 2002. My grandmother saw her for the first and, unfortunately, the last time as well in October. When we visited her, the entire visit was very strange, not at all like usual, and I sobbed all the way home without being able to explain why.

"On December 11th I knew why, as she had now died. The last visit had been so different from usual. My grandma had sat in the kitchen and given me such a strange look.

"Very strange things have been happening since her death. She appears in dreams that seem so real I think I hear her voice. Many other things have happened as well, but I do not want to write it down as no one would believe me.

"But I will say one thing: I always fall asleep in front of the TV. I woke up one night and the TV was off. I suddenly saw something white, like weak smoke, moving toward my daughter's room. I was terribly afraid."

The Dialog, Part 4:
Our Dual Ego

"Chance is a nickname for providence."
(Nicolas-Sébastien de Chamfort, French Author, 1741-1794)

Al rolled his eyes: "I am slowly beginning to realize where you are headed with this!"

"I just mean that physics is facing its biggest challenge," Zach calmly replied.

"And what would that be?"

"Demonstrating that quantum physics includes large objects as well – whether animate or inanimate."

Al waves the idea aside: "Oh great – that is why we are so good at moving through walls."

"Hang on, don't go dragging everything through the mud. Of course macroscopic objects can't tunnel. But, look at yourself from a microscopic point of view instead. What is your composition from a biological, chemical and physical perspective?" Zach gives his friend a questioning look.

"Just like a human is supposed to be composed: of organs, bones, cells, etc. What else do you want to know?"

"That is the macroscopic level of biology. So what about your chemical make up?"

"Of course, if I take a closer look at my own microcosm, I will inevitably come across macromolecules like protein chains, for example."

"But that is just the tip of the iceberg."

"Yes. Because even they, as I still remember from Chemistry class, are also composed of smaller units, which in turn, are composed of atoms and other elementary particles. The lowest level can, once again, be attributed to physics."

Zach leans forward with a challenging look on his face. "Exactly, and they, in turn, can be described by their wave function."

"Hang on a second! This wave-like nature would have collapsed long before larger and larger molecular units could combine to form even a single cell," Al disagrees.

Zach shakes his head. "But that is the big question. Is it not true that an individual molecule or a tiny particle in you fails to make sense until it communicates with other particles in a subtle way in which the end result is good old Al standing in front of me, with all of his knowledge and his positive and negative characteristics that make him an individual?"

"I didn't question that!"

"Good, but do you think the sum of all wave functions is a living individual or just a mere zero-sum situation?"

Al laughs. "What else would it be?"

"My dear Al, I am now going to postulate that the sum of these wave functions is your dual ego – your soul or some sort of super wave is to the body what the wave is to the particle."

"My dear Zach," Al slowly replies, "That, truly, is mere speculation."

"Yes, but a very helpful one. Take the human brain, for example. It also consists of molecules, atoms, electrons and nuclei. Why shouldn't we be able to apply wave mechanics to it? And behind the brain we find the spirit. Are you suggesting it is material?"

Father Predicts His Death

Before continuing the discussion, I would like to use this opportunity to introduce a few more phenomenal examples and the backgrounds behind them. Mr. Lothar Decker of Hamburg, who is currently working on a publication on synchronicity, conveyed the following story to me:

"My old father was bed-ridden by pleurisy from which he would likely not recover. He seemed to sense that the time of his parting had come. The reconciliation between the two of us had lifted a heavy burden from his soul.

"I had a strange dream one night during those days; days that moved between hope and despair in regard to his state of health. My father, who was wearing pajamas and a bathrobe, was about to board an omnibus that was ready to depart, but missed it. The doors closed right in front of his nose and the vehicle left without him.

"The symbolism in the dream was revealed to me the next night in which my father actually passed away. The dream was an allusion to his impending death, showing his last trip – the journey to the other side in an "omnibus" (which is actually Latin for "for everyone") – that had not yet occurred, but which was directly before him. The clothing itself was a reference to the time of his passing. My father had departed in the early hours of the morning, a time between night (pajamas) and day (bathrobe).

"A few hours after my father's death, a friend I was living with at the time told me his clock had stopped at 3:35 a.m. even though he had just replaced the old batteries.

"I drove to my father's house, bade him farewell and returned to my friend. He told me he didn't have to change the batteries after all, as when he went to do so, the clock was working again.

"It was not until later that I found out that, according to the death certificate, my father had died at 3:30 a.m., but I already knew that as he had let me know the exact time (3:35 a.m.) by means of my friend's clock."

The Scientist's Secret Fear of the Unknown

As mentioned before, I am not trying to create a scientific platform for mysticism. On the other hand, subjects like "paranormal perceptions", "telepathy", "prophetic dreams", etc. could be moved into a completely new dimension using quantum mechanics.

My first thoughts in this direction began back in 1990 and were not influenced by literature on the topic. At the time – independent of the correspondence between C. G. Jung and Wolfgang Pauli (who was unknown to me at the time) – I was unaware such literature even existed.

Also at that time, I was working as the editor of a division for "Research and Development" at the magazine *highTech* in Munich. One day over coffee, I began a discussion with Dr. Reinhard L., a new colleague who had joined our editorial team right after the turn of events in the former GDR. Reinhard had previously studied physics at the Russian Universities in St. Petersburg (which was still Leningrad back then) and Yerevan.

Since I am a chemist, I wanted to know whether he, as a physicist, thought the idea of a human "quantum spirit", which could easily explain such "supernatural phenomena", violated any laws of physics. Reinhard said no and told me that in Russia high-ranked scientists had already begun contemplating the subject.

Still something kept me from openly expressing my opinions on the topic. As a scientist and journalist, the fear of being tagged an esoteric was too great. I now realize that was a mistake. In fact, ambassadors from the academic community encouraged me to write this book. These academics include researcher Dr. Markolf H. Niemz of Heidelberg. In his book *Lucy with C*, he proposes the thesis that when a human dies, his soul departs at the speed of light. A big step for a thoroughbred academic!

Niemz teaches medical engineering at the University of Heidelberg. He also focuses on near-death research. The latter gave him the idea for his revolutionary thesis.

During a so-called "near-death experience", the party in question suddenly has the sensation that his soul has separated itself from his physical body and is hovering above the stage on which the events are taking place. He basically watches his body from above, from a distance, lying on a hospital bed, for example. He can also "watch" as doctors and nurses fight for his life.

Just moments later a tunnel seems to open. The person in question often feels "pulled toward" the long tunnel and moves toward a bright, but not blinding light at the end of the tunnel. This type of near-death experience is the most commonly described one to date. One may even have the sensation of seeing one's life pass right before his eyes right before seeing the tunnel to the "other side". The person in question often sees important scenes from his life "pass right in front of his spiritual eyes". This is usually in the form of images, a sort of spiritual movie in which he is a disengaged bystander, watching himself from a removed perspective.

Niemz sees parallels between this "light at the end of the tunnel" and a simulated journey in a space ship traveling close to the speed of light. The so-called "search light effect" creates the impression that everything in front of the observer is moving toward him. A similar effect can be found when driving through a snow flurry in winter. In this case, it seems as if bundles of snow are actually coming at the driver from the front, when in reality, the snow is falling almost completely vertical, even if there is a slight wind. Similarly the result of this effect on an object moving through the universe at almost the speed of light would be the sensation that there were a bundle of light rays coming toward it from in front, while the rest of the universe would seem to get darker the closer the object got to the speed of light. Once 95% of the speed of light had been reached, the observer would have the feeling of heading toward a dark tunnel toward a shining source of light, i.e. a light at the end of the fictive tunnel.

Are we really looking at the human soul leaving the body at the speed of light as Niemz suggests? Or are we, instead, being fooled by the opiates, endorphins and enkephalins the body emits in situations of extreme stress? Just as marathon runners have a sensation of being extremely happy, some scientists believe it is also completely natural for people, who are drowning or freezing to death, to recount the same euphoric state shortly after being rescued. This theory sounds quite plausible and is also quite obvious. What remains to be explained, however, is why the stories told by the parties in question are so very similar. Finally, there is still no proof that this effect is the result of the high concentrations of endorphins that are at work in the body as this type of evidence is likely very difficult to collect. In life or death situations, the acting physicians have very different priorities.

One weakness I see in Prof. Niemz's theory is that although a human undergoing a near-death experience is fighting for his life, he must technically still be alive - even if his heart stops - as we are able to reverse the process of dying, yet we are not able to bring the dead back to life. The posit of a soul departing at the speed of light and "perceiving" a light at the end of a tunnel as a result of the search light effect, would have to have already crossed that threshold. At that speed, the soul would already have crossed the moon's path in 1.3 seconds and would have already reached the planet Mars' orbit by the time the person was revived, a process that must take place within a matter of minutes. How on Earth is the poor soul supposed to find its way back into the body under those kinds of conditions? That is a notion that is a bit hard to digest! Yet, I do give Prof. Niemz credit for having the courage to open the door to a completely new dimension in wave mechanics and for encouraging (hopefully) natural scientific research to enter into new terrain. This latter point is especially important to me and I will address it in more detail later on.

Is the Other Side the Great Internet of Reality?

I mentioned Prof. Dr. Hans-Peter Dürr, former head of the Max Planck Institute for Physics in Munich, very briefly in an earlier chapter. Today, Dürr represents the opinion that the dualism of the smallest particles is not limited to the subatomic world, but instead is omnipresent. In other words: the dualism between the body and the soul is just as real to him as "wave-particle dualism" of the smallest particles. According to his view, a universal quantum code exists that applies for all living and dead matter. This quantum code supposedly spans the entire cosmos.

Consequently, Dürr believes – again based on purely physical considerations – in an existence after death. He explains this as follows in an interview he gave (see "Further Reading"):

"What we consider the here and now, this world, it is actually just the material level that is comprehensible. The beyond is an infinite reality that is much bigger. Which this world is rooted in. In this way, our lives in this plane of existence are encompassed, surrounded, by the afterworld already. When planning I imagine that I have written my existence in this world on a sort of hard drive on the tangible (the brain), that I have also transferred this data onto the spiritual quantum field, then I could say that when I die, I do not lose this information, this consciousness. The body dies but the spiritual quantum field continues. In this way, I am immortal."

Dr. Christian Hellweg is also convinced the spirit has a quantum state. Following his studies in physics and medicine, he researched brain function at the Max Planck Institute for Biophysical Chemistry in Göttingen for many years. He was able to show that information in the central nervous system can be phase encoded. In recent years, he has dedicated himself to studying the body-soul issue and researching phantom perceptions and hallucinations. He is especially interested in tinnitus, a phantom perception in the sense of hearing. He has also specialized in the therapy thereof. He summarizes his thesis as follows:

"Our thoughts, our will, our consciousness and our feelings show properties that could be referred to as spiritual properties…No direct interaction with the known fundamental forces of natural science, such as gravitation, electromagnetic forces, etc. can be detected in the spiritual. On the other hand, however, these spiritual properties correspond exactly to the characteristics that distinguish the extremely puzzling and wondrous phenomena in the quantum world. Quantum world, in this case, refers to that realm of our world that is not yet factual; in other words, the realm of possibility, the realm of uncertainty, where we do "know what", but do not exactly "know when or how". Based on the context of traditional physics, it can, out of necessity, be concluded that this realm must actually exist."

American physicist John Archibald Wheeler hits a similar nerve, "Many scientists hoped…that the world, in a certain sense, was traditional - or at least free of curiosities such as large objects being in the same place at the same time. But such hopes were crushed by a series of new experiments." There are now university research teams examining the interaction between consciousness and material. One of the leading researchers in this field is physicist Professor Robert Jahn of Princeton University in New Jersey. He concludes that if effects and information can be exchanged in both directions between the human consciousness and the physical environment, then one must also assume a resonance or "molecular binding potential" for the consciousness as well. In summary: according to this theory, one would have to award the consciousness the known quantum properties as well. In his opinion it makes no sense, to assign terms such as information or resonance to either the physical world or the spiritual consciousness or to separate physical effects from spiritual effects.

Quantum physicist David Bohm, a student and friend of Albert Einstein, made similar claims. His summary: "The results of modern natural sciences only make sense if we assume an inner, uniform, transcendent reality that is based on all external data and facts. The very depth of human consciousness is one of them."

Nuclear physicist and molecular biologist Jeremy Hayward of Cambridge University makes no secret of his convictions either: "Many scientists who are still part of the scientific mainstream are no longer afraid to openly state that consciousness / awareness could, in addition to space, time, matter and energy, be a fundamental element of the world – perhaps even more fundamental than space and time. It may be a mistake to ban the spirit from nature." It is even questioned as to whether matter should be considered a fundamental element of the universe. At least Hungarian physicist and musician Ervin László is convinced that "an immaterial ocean of energy fills cosmic space".

The Dialog, Part 5: Where Is God?

"The first drink from the cup of natural science brings atheism, but at the bottom of the cup waits God." (Werner Heisenberg)

Al makes a dismissive hand movement: "I can see I won't be able to put you in an esoteric box. There are just too many renowned personalities who back theses that used to be considered parascience."

Zach nods in agreement, "Exactly, but my point isn't to provide scientific arguments for so-called ghost whisperers. Rather, I am of the opinion that certain phenomena should be examined from a scientific point of view once and for all, free of prejudice and modern wave mechanics provides the perfect platform."

Al leans in, "Let me ask you something completely unrelated. Do you happen to be an atheist?"

Zach shakes his head, "No, what gave you that idea?"

"Because you seem to be close to people who want to explain everything, and I mean everything, using science. But there is no place for God in that view of the world."

"Quite the contrary, dear Al. I believe God created not only the Earth, but the entire universe. The actual act of creation took place around 13.7 billion years ago, long before the Earth was created by the so-called big bang. The current universe was practically born of a single element during the event."

"I am aware of the latter," Al comments slightly insulted.

"Of course, but have you thought about the consequences?"

"I am not sure what you are thinking about right now. The big-bang theory has been convincingly backed by science."

"I am thinking about the inconceivable amount of information that one secretive element must have contained right before the big bang. Just as egg or sperm cells contain an enormous pool of information that undeniably influences the life that results when they are combined, this predecessor to the universe must have had all the data that now makes up the current universe."

"An interesting thought. Based on that idea, the explosive causing the explosion that created the universe was thus information, right?" Al looks questioningly at Zach.

Zach thinks about that for a moment, leans back and with intentional emphasis responds, "In the beginning was the Word and the Word was with God and the Word was God. That is what it says in the book of John."

Al swallows audibly. "I have to admit, discussions with you are always interesting. But do you really believe it?"

"Yes, because natural science and religion are no longer in opposition to one another. Instead, they complement one another like someone's left shoe and their right shoe. Together, they form a perfect pair."

Before continuing the discussion, I would like to relate two other events that also come from absolutely trustworthy sources.

The Soldier who Said Goodbye

Sigrid Donath from S. recollects, "In 1943 the bomb attacks on my hometown of Cologne got worse and worse. I spent many nights in the air-raid shelter with my mother while my father was stationed in Italy as a member of the German Armed Forces. As the situation became more critical, we accepted an invitation from relatives in the Harz mountains and left Cologne with heavy hearts.

"In the spring of 1945, I had just turned 17, I had a strange encounter in my new surroundings. One day when I went into the cellar, I suddenly saw my father in front of me, dressed in his military uniform. I was frightened and spoke to him. He didn't react and disappeared all at a sudden, when I turned on the light. I screamed as I ran up the stairs and told my mother and my relatives about the occurrence. They immediately searched the cellar without result.

At the end of the war my father was considered missing for many years. Not until much later did we receive the sad assurance that he had fallen in the spring of 1945 at the Ghedi airport, close to Milan in Northern Italy. Today, I am fairly certain my father had wanted to say goodbye to me at the time."

"Willy – Where on Earth Did You Come From?"

Another story that was credibly conveyed to me is along the same lines. It is a story that has been told in our family for generations. Dorette Stahl of Altenau (one of the author's great-great-grandmothers) had three children named Willy, Wilhelm and Louise. During the First World War, Dorette suddenly saw her son Willy go through the living room wearing a bandage on his head. "Willy – where on Earth did you come from?" Her loud yell caused the remaining family members to come running, but the apparition disappeared as quickly as it had appeared. A short time later the family received news that Willy had been injured the very same day he had appeared to his mother in the living room.

The Dialog, Part 6:
Why There Can Be No Stable Channels of Communication on the Other Side

"There is no matter as such. There is only the exhilarating, invisible spirit as the original cause of matter with the secretive Creator, who I am not afraid to call God." (Max Planck)

"We now have enough examples of higher-order coincidences, although the latter examples dealt more with apparitions, like the case of Max Planck scientist James G. Are you, dear Zach, able to explain them using the laws of wave mechanics?"

"I don't think science has come far enough for me to be able to answer that question with a 'Yes'. It is highly probable that ever since the big bang, large parts of the cosmos are connected, as if by an invisible band. That is the principle of entanglement, which has now been confirmed by experiment."

"Okay. But do you really believe that the phenomena that seem so much like ghost stories really exist?"

"What other explanation is there? The named sources are absolutely reliable!"

"But how is it all supposed to work?"

"Well, good old Albert Einstein himself mentioned spooky action at a distance in speaking of entanglement. And I could quote Prof. Dürr at this point again too. He believes in a quantum code that extends across the entire cosmos. Accordingly, that which we perceive in our everyday lives is just a part of reality."

"And how can we find out about the other part of reality?"

"By pure intuition, through perceptions, spontaneous intuition and feelings we cannot explain based on our usual experience."

Al keeps at it. "You really think Mrs. Donath and Mr. Grant saw spirits and that those spirits actually existed?"

"Yes. And no one got to the point better than Prof. Dürr when he said that our current life is already surrounded by the other side. However, I struggle with the word spirit. It is too closely related to ghosts and is misleading. This type of communication materializes from the quantum state of the soul and that state is just as real to me as the physical body."

"Okay, Zach, if that is what you truly believe then answer one last question."

"And that would be?"

"Let's take a look to the future. When do you think we will be able to communicate with the other side directly and reliable through suitable information channels?"

"There probably never will be a stable form of communication with the other side. Instead, the contact will always be pure coincidence."

"Aha! And the great preventer is Werner Heisenberg, right?"

"Dear Al, although the Heisenberg uncertainty principle probably does play an important role, we still have to wait and see whether its grid isn't too coarse. Maybe someday a clever inventor will present a Heisenberg compensator. But I don't think so and I certainly hope not."

"Why do you hope not? A reliable form of communication with the other side would certainly be associated with an enormous knowledge gain!"

"You are looking at it far too rationally. But think about the consequences for a minute. Billions of living people would thirst for the chance to communicate with dead relatives. No, that just can't be allowed to happen. A permanent line to the other side would make us hopelessly addicted!!! I am also convinced that God created a rule that governs the uncertainty principle, to prevent communication with the other side from becoming a form of tourism. However, he did leave a back door open for sporadic contact."

"That sounds pretty plausible. What else do you have?"

"The entanglement principle in wave mechanics can also be used to explain the so-called seventh sense that animals have."

"Interestingly enough, biochemist Rupert Sheldrake of Cambridge has been intensively involved in researching that topic recently. Do you know what his theory of morphogenetic fields actually reveals?"

"Essentially, Sheldrake is on the right track. In his opinion, however, there is no need to pull a new field theory out of a hat. Instead, all phenomena can be easily explained using wave mechanics. The entanglement principle provides a theoretical basis for spiritual healing, as well as for the phenomena of telepathy and remote viewing."

"The latter truly is completely esoteric."

"I am familiar with a case that is not esoteric at all. Once again, the story is about a scientist. Let's take a look at this story and then discuss the various conclusions that can be drawn from it and that pertain to evolution as a whole."

"Seeing" a Major Fire from 450 Kilometers Away

One of the most spectacular and yet best documented cases of remote viewing can be found in a report by Swedish natural scientist and theologian Emanuel von Swedenborg (1688-1772). Dr. Hans Schwarz, professor of Evangelical theology at the University of Regensburg, studied contemporary sources related to the matter. He summarizes Swedenborg's, who was originally from Stockholm, experience in his book *Wir Werden Weiterleben* (we will live on):

"One evening in July of 1759, Swedenborg had been invited to dinner in Göteborg, a city approximately 450 kilometers southwest of Stockholm. Swedenborg suddenly paled and became visibly agitated. He went out into the garden and returned with the message that a major fire had broken out in Stockholm, not far from his own house. He claimed that the fire was moving quickly and was worried about his manuscripts. He finally calmed down and sighed in relief saying, 'Thank God! The fire went out three doors down from my house.' As a few of the guests that were present had houses or friends in Stockholm they were naturally quite upset as well. One of them told the governor about the event that same night. The next day Swedenborg shared details of the fire. The message of the alleged disaster spread quickly throughout the city. But it was not until the day after that a messanger came from Stockholm confirming every detail of the story."

From the Primordial Soup to DNA

A few scientists claim that evolution and the creation of life on Earth are mere coincidence. Accordingly, it is highly probable that the Earth is the only inhabited planet in the entire universe. Why?

Well, mathematicians have calculated that, should that be true, life and all of the complex biomolecules it encompasses came into existence under a probability of 1 to 1040,0000. That is a 1 followed by 40,000 zeros. Just for comparison: if you were to cover the Earth's surface with 1-Euro coins, stacking them 1 meter high, and one of them were marked in red, you would have to find that one red coin on your very first try.

I am convinced that evolution was not a coincidence, but the result of a cosmic quantum phenomenon, which in religious terms is often referred to as the act of creation. In this next section, let us take a little journey through time and examine the creation of life:

As far as we can tell today, the big bang occurred 14.7 ± 0.2 billion years ago. Our solar system did not come into existence until about nine billion years later. At that time, the universe already had the structure we see today with billions of galaxies consisting of billions of stars. The genesis of these structures was determined by a mysterious quantum code in which everything was connected via the entanglement principle and thus independent of the distances between them.

The world must have been a strange site four-and-a-half billion years before our time. The sunlight we are so familiar with did not exist nor did the Earth, moon or planets. Instead, there was a massive cloud of gas and dust floating through space. It was much denser at its core than at the outer areas and invisible heat waves were emitted from its center. This state must have lasted a few thousand years when a visible light suddenly penetrated the dark. A new-born star shone through the dust cloud. That star was our sun.

A look behind the curtains of this cosmic stage shows us that elementary hydrogen, the simplest and lightest of all chemical elements, was the main actor. The pressure and temperature within the cloud had risen to such an extent that the core began to melt, a process also referred to as fusion. The process is still ongoing on the sun and without it there would be no light on Earth, no warmth and thus no life. It was the first act in the creation story. The process created a world reigned by physics alone. It was the *era of physical evolution.*

As the cosmic stage progressed, the shroud of mist continued to clear. Gigantic beds of fog crashed into the young sun, as if to fuel the newly ignited fire. Other parts defied the pull of gravity and orbited on elliptical paths. This part of the ylem or primordial substance initially orbited in a series of interlacing paths in which collisions were common. Modern model calculations confirm that the continuous collisions must have resulted in an assimilation of the paths. The prime fathers of the planets we see today came from those cluster-like conglomerate microplanets. The resulting planets soon began to cool more and more from the outside in. A hard crust formed on their surface – the primeval continents had been born.

Even 700 million years after the sun had sent its first ray of light out into the universe, things on Earth were still rather uncomfortable: the primeval oceans, lukewarm brews of salt that reeked of sulfur and ammonia, surged against bizarre continents of volcanic basalt. Wind, weather and frequent volcanic explosions continuously reformed the surface. Yet, neither the chirping of a cricket nor the cry of a bird could be heard in the air – the Earth was still uninhabited. It was the second act of the creation story – the *era of chemical evolution.*

In 1953, a 23-year old chemist named Stanley L. Miller of the University of Chicago was able to establish evidence that the next phase was indeed a purely chemical evolution. He was able to do so by simulating the environmental conditions that exist in primeval Earth in a lab experiment. To recreate the conditions, he mixed a "primeval atmosphere" consisting of methane, ammonia and hydrogen gases together in a receptacle in the presence of water. He then fed energy into the mix using electrical charges. This simulated lighting on primeval Earth. Sometime later, after the cooling, he was able to prove the existence of numerous organic compounds that had been formed during the process, including fatty acids, carbohydrates and, particularly important, a few amino acids. These are the components that make up proteins (protein molecules), which also include the enzymes that are decisive in controlling all the biochemical reactions that occur in living beings. Later experiments under similar conditions also produced purine and pyrimidine; important components of the two nucleic acids ribonucleic acid (RNA) and deoxyribonucleic acid (DNA), the inheritant material found in all living beings which has the ability to replicate itself. Thus, it is probable that this or similar compounds were created during the early phases of the Earth. This, in turn, proves the creation of life from chemical conditions.

The following experiments, on the other hand, prove that the inorganic substances found in oceans and silts containing zinc, in particular, played an important mediating role in the process. These substances are able to separate the wheat from the chaff in regard to the wide variety of amino acids that are chemically possible. As a result, the silt mainly served as a catalyst for the synthesis of the amino acids that are essential for life to exist. Accordingly, oceans could be referred to as the "primeval soup of life". Subsequent and targeted condensation reactions were able to produce increasingly complex molecules until the creation of the first self-replicating unit. This was the third act of the creation story – the *era of biological evolution.*

Life already existed three billion years ago. There are fossils – so-called stromatolites – that prove the existence of microscopically small, bacteria-like organisms. Although some fossils of bacteria-like organizations have been found that are even older (about 3.5 billion years), scientists have been unable to determine their exact classification with complete certainty. Thus, life must have originated prior to that, perhaps 3.7 to 3.8 billion years ago. Even though the exact process through which life was created is unclear, all theories agree that molecules with the ability to self-replicate, meaning that under certain conditions they are able to create identical copies of themselves, were created during the process and that these molecules could generally be referred to as genes. Such molecules represent one of the requirements for the development of life, because they ensure that properties are passed on from generation to generation and thus enable reproduction.

Thus, it should come as no surprise that everything found in nature today is ultimately chemistry and everything, whether living or non-living, is composed of the same chemical elements. Even if we are able to explore the expanses of the cosmos using space probes and telescopes, we will always run into the same chemistry. Whether the red desert terrain on Mars, whose color is the result of iron oxide, or the fantastic colors of the Jupiter's clouds in which mainly unexplored organic molecules roam, one thing holds true: chemistry is everywhere. A few years ago when the Hubble telescope circling the Earth discovered an inferno of gas and dust in the Orion Nebula, one that was similar to the state of our own solar system 4.5 billion years ago, humans were finally able to not only see the depths of the universe, but for the first time ever, catch a glimpse of their past as well. If, after 700 or 800 million years in that chaos, planets actually will begin to form, the chemistry involved in that process will be the same chemistry that allowed similar molecules to form on primeval Earth when the conditions are favorable.

The Dialog, Part 7:
A Series of Conclusions

"God does not play dice."(Albert Einstein)

Al dismisses it with a wave of his hand. "I also don't believe that evolution was a mere coincidence. I consider the idea that among the billions of galaxies, which themselves have billions of suns, there is supposedly only one single inhabitable planet, to be a delusion as to the center of the cosmos."

Zach agrees, "I find that easy to believe. You, in your designer glasses and all, certainly don't look like a medieval pope. But if evolution wasn't a coincidence, it must be the result of an ordering principle prescribed by nature. And which could that be?"

Al looks at him over the top of his glasses, "Ever heard of someone by the name of Charles Darwin?"

"Of course, Charles Darwin developed the theory of a natural principle of evolution through gradual variation and natural selection. His theory explains the slow division of organisms into a wide variety of types as a result of their adaptation to their environment. But Darwin's principle can't explain the actual core of evolution, that of the development of the first gene from the simple contents of the primeval soup."

"That was just a form of chemical evolution."

"True, but that could not have been mere chance. Instead, I would like to compare and contrast Darwin's principle of natural selection to a quantum principle of chemical and biological evolution."

Al laughs and raises his glass to his friend, amused. "Please tell me you're not about to go after Darwin too. Let's have a drink before you do. How, in your opinion, is that supposed to work?"

Zach savors a sip of wine, letting it rest on his tongue. "Dear Al, the foundation of life is molecular. Molecules are quantum systems and exist in quantum states. And there is an universal principle of organization that governs the cosmic code, a principle that eventually led to evolution. In other words, the core of evolution is based not on the natural selection of genes, but on the wave functions of quantum states, which with all of the information stored in them were first materialized in the first genes."

"An interesting theory..."

"...that can be used to easily explain a whole series of phenomena, because it is a holistic, in the truest sense of the word, universal approach."

Al nods acceptingly. "Then space would be populated by extraterrestrial intelligence, would it not?"

"Inevitably, yes, because the quantum code and the principles of evolution are just as applicable in the Adromeda nebula as in our Milky Way or the furthest insular world in the universe. This naturally assumes that a planet is within the so-called life zone of a star."

"You mentioned other phenomena that could be explained by that theory as well."

"When I vacation on tropical areas I like to go diving in the coral reef. I have always asked myself why the individual members of a school of fish basically react in a synchronized manner when interrupted and are able to change directions within a fraction of a second. I think the flocking behavior of small animals involves quantum effects as well."

"Then the same would have to be true for ant farms and beehives."

"Thank you for bringing that up, dear Al. I read an interesting article on beehives in the newspaper *Die Welt*. According to the latest research, we can understand the organization of the hive much better if we look at the bee colony as a super organism composed of 30,000 to 50,000 bees."

"I find that very interesting. I would love a copy of that article if you can get me one."

"I'd be pleased to. But, in closing, let me come back to Prof. Dürr. He sees the wave mechanic form of quantum physics as the key to immortality."

Al takes a deep breath, "I find the idea very comforting. But, are there any indications that the theory could be true, besides the case you described in which people had experiences with the other side?"

"Yes, one I see is the example of phantom pain, a pain that occurs when someone loses a body part. Even though the patient no longer has the body part in question, he still feels pain from time to time. Traditional medicine, on the other hand, interprets the phenomenon as nerves that have been damaged during the loss of the body part, nerves that are responsible for aspects such as communicating information about pain in the amputated organ. The brain then interprets the signals as pain in the amputated body part."

"That sounds logical, doesn't it?"

"Not at all, otherwise the pain would be permanent. Moreover, our brain is programmed to eliminate false signals."

Al glances at his watch, "You mean that the signals are actually real?"

"Yes, because the soul does not allow amputation."

Al makes a regretful gesture. "It has gotten pretty late. I have to go, but we should continue our discussion sometime soon. It was really very fascinating. I am sure I will be thinking about the aspects of the physics of immortality for the next few days."

"Then I would encourage you to think about the death of the universe as well."

"Dear Zach, I don't have to think about that at all. The universe will not be able to prevent a thermal death. However, when that will occur is still unknown."

"I doubt that the universe will die from Big Chill."

"Then you must not have been paying attention in thermodynamics. Just as wind can turn a sandcastle back into a dune, but never vice-versa, so can entropy, as a dimension of chaos, remain the same or increase in a closed system, but never decrease. Thus, once all possible processes have been exhausted and there is one constant temperature and maximum entropy has been reached, every possible evolution within the system will have to come to an end. Period."

"First of all, Al, it is not clear whether the universe can even be viewed as a closed system. On the other hand, I am convinced that the quantum properties of the universe can compensate for the increase in entropy and can thus prevent thermal death from occurring. Just as humans and animals are immortal, so must the universe be immortal as well. When God created the Universe, he did not assign it an expiration date. Period."

Zach shakes his friend's hand heartily upon his departure. "You were a great guest. Thank you for the very successful evening."

"Time will tell whether you are right."

At the door, Zach pats Al on the shoulder. "I don't know if I will live to see that. But I certainly hope that international research in the field will move out of shadows and into the limelight. In regard to the results, I will just have to wait and see, because the right discovery always finds a way when the time is right."

Further Reading

Al Khalili, Jim: Quantum: A Guide for the Perplexed. Weidenfeld & Nicolson, 2004.
Atmanspacher, H., Primas, H., Wertenschlag-Birkhäuser, E. (Hrsg.): Der Pauli-Jung- Dialog und seine Bedeutung für die moderne Wissenschaft. Springer-Verlag Berlin Heidelberg 1995.
Becker, V.J.: Gottes geheime Gedanken. Books on Demand GmbH, 2006.
Chown, Marcus: Warum Gott doch würfelt. Deutscher Taschenbuch Verlag, 2005.
Cox, Brian; Forshaw, Jeff: The Quantum Universe (And Why Anything That Can Happen, Does). Da Capo Press; 1st edition, 2012.
Froböse, Gabriele; Froböse, Rolf: Lust und Liebe – alles nur Chemie? Wiley-VCH Verlag, 2004
Froböse, Rolf: Die Seele existiert auch nach dem Tod. Die WELT vom 25. April 2008.
Froböse, Rolf: Wenn Frösche vom Himmel fallen. Die verrücktesten Naturphänomene. Wiley-VCH Verlag, 2007.
Froböse, Rolf: Mein Auto repariert sich selbst. Und andere Technologien von Übermorgen. Wiley-VCH Verlag, 2004.
Gell-Mann, Murray: Das Quark und der Jaguar. R. Piper Gmbh & Co. KG, 1994.
Goa, Shan: Understanding Quantum Physics: An Advanced Guide for the Perplexed. Amazon Kindle Direct Publishing, 2011.
Green, Brian: The Hidden Reality: Parallel Universes and the Deep Laws of the Cosmos. Vintage; Reprint edition, 2011.
Gribbin, John: Schrödingers Kätzchen und die Suche nach der Wirklichkeit. S. Fischer Verlag GmbH, 1996.

Hood, Bruce: The Supernatural Sense. A Scientist Explains Why we Believe in Intuition, Superstitions, and God. HarperCollinsPublishers, San Francisco, 2008.
Hopcke, Robert H.: Zufälle gibt es nicht. Die verborgene Ordnung unseres Lebens. Deutscher Taschenbuch Verlag, 2001.
Klein, Stefan: Alles Zufall. Die Kraft, die unser Leben bestimmt. Rowohlt Taschenbuch Verlag, 2006.
Kumar, Manjit: Quantum: Einstein, Bohr, and the Great Debate about the Nature of Reality. W. W. Norton & Company; Reprint edition, 2011.
Lambeck, Martin: Irrt die Physik? Verlag C. H. Beck, 2003.
Lesch, Harald: Quantenmechanik für die Westentasche. Piper, 2007.
Mardorf, Elisabeth: Das kann doch kein Zufall sein. Verblüffende Ereignisse und geheimnisvolle Fügungen in unserem Leben. Kösel-Verlag GmbH & Co. KG 2002.
Meier, C.A.(Hrsg.): Wolfgang Pauli und C.G. Jung. Ein Briefwechsel 1932 - 1958. Springer-Verlag Berlin Heidelberg 1992.
Niemz, Markolf H.: Lucy mit c. Mit Lichtgeschwindigkeit ins Jenseits. Edition BOD, 2005.
Rosenblum, Bruce; Kuttner, Fred: Quantum Enigma: Physics Encounters Consciousness. Oxford University Press, 2011.
Röthlein, Brigitte: Die Quantenrevolution. Deutscher Taschenbuch Verlag, 2004.
Thomas, Andrew, H.: Hidden In Plain Sight: The simple link between relativity and quantum mechanics. CreateSpace Independent Publishing Platform, 2012.
Vaughan, Alan: Incredible Coincidence. The Baffling World of Synchronicity. Ballantine Books, New York, 1989.
Zeilinger, Anton: Einsteins Spuk. Teleportation und weitere Mysterien der Quantenphysik. Goldmann, 2007.
Zeilinger, Anton: Einsteins Schleier. Die neue Welt der Quantenphysik. Goldmann, 2005.

Photo credits:

Cover image: The Horsehead Nebula, located about 1,500 light-years from Earth, is a dark cloud in the constellation Orion and is also one of the best known astronomical objects. It is assumed that new stars and predecessors to planets are created within such dark clouds. Since evolution was not a coincidence, some of them may even be able to support life at some point in the future. (Source: NASA)

CPSIA information can be obtained at www.ICGtesting.com
Printed in the USA
BVOW08s0506010715

406952BV00002B/109/P